Hanley Sylvanus, Theobald William

Conchologia Indica

Illustrations of the Land and Freshwater Shells of British India

Hanley Sylvanus, Theobald William

Conchologia Indica
Illustrations of the Land and Freshwater Shells of British India

ISBN/EAN: 9783743319264

Manufactured in Europe, USA, Canada, Australia, Japa

Cover: Foto ©berggeist007 / pixelio.de

Manufactured and distributed by brebook publishing software
(www.brebook.com)

Hanley Sylvanus, Theobald William

Conchologia Indica

PREFACE.

NOT because it is complete, not because we could not render it much more complete, but because the stimulus given to collectors by our iconography has added such a multitude of species to the conchological fauna of British India as to render it impossible to keep pace with modern discoveries, is the reason that we have resolved to terminate a publication whose anticipated limits have already been exceeded. We never either intended, or hoped, to publish an exhaustive work; our aim was merely to assemble in one book species which were dispersed in scores of other volumes. We have shown what has been done up to 1873, and thus facilitated the labours of those who may aspire to produce a more perfect conchology. It is something to have caused the delineation of types before they were lost, forgotten, or destroyed; and the search for the actual specimens originally described by Benson, Pfeiffer, Hutton, Blanford, &c. has consumed both time and money.

Our best thanks are due to Major Sankey (the executor of our late dear friend W. Benson), to Major Godwin-Austen, Major Hutton, the late Frederic Layard, and last, not least, to W. Blanford: their friendly aid has enabled us to rightly determine and illustrate many a puzzling species.

The Editors regret that the figures of some of the more minute shells are not so well executed as they expected; but lithography is scarcely compatible with sharp definition, and to correctly magnify such little objects strains the eyesight to an extent few artists will dare to venture.

It is important to observe that the inclusion of a species in the systematic list which follows by no means indicates that the Editors recognise its distinctive merits; their aim has been mere elucidation, and friendship forbids all criticism (however tacit) which can possibly be avoided. The list itself is not put forth as a scientific arrangement, but only as a convenient sequence.

After an interval of two or three years it is hoped that materials for a Supplement (the malacological portion of which will be edited by Major Godwin-Austen) will be accumulated, and to further that design collectors are earnestly requested to forward specimens to the English Editor, to the care of his publishers, Reeve and Co.

SYSTEMATIC LIST OF SPECIES

PUBLISHED UP TO 1874.[1]

BIVALVES.

Taxisiphon.

rivalis, Bn. t. 116, f. 1, 4.

Novaculina.

Gangetica, Bn. t. 116, f. 7 : var.? f. 10.

Pisidium.

Clarkianum, Nev. t. 155, f. 9.

Cyclas.

Indica, D. t. 155, f. 10.
Avana,[2] Th. (Ava).

Corbicula.[3]

Bengalensis, Desh. t. 155, f. 6.
Bensoni, D. t. 138, f. 1, 4.
Cashmirensis, D. t. 138, f. 2, 3.
Invadiens, Bl. t. 155, f. 7.
occidens, D. t. 138, f. 8, 9.

regularis, Prime, t. 138, f. 5, 6.
striatella, D. t. 138, f. 7, 10.
trigona, D. t. 155, f. 7.

Scaphula.

celox, Bn. t. 116, f. 8, 9.
Deltae, Bl. t. 116, f. 2, 3.
pinna, Bn. t. 116, f. 5, 6.

Mycetopus.

Bensonianus,[4] Lea, t. 9, f. 1 (as Soleniformis).

Trigonodon.

crebristriatum, t. 9, f. 3, 5.

Pseudodon.

Ava,[5] Th. (Ava).
inoscularis, Gould, t. 9, f. 3.

Unio.[6]

Birmanus, Bl. t. 42, f. 1.
Bhaumensis, Th. t. 155, f. 2.
Bonneandi, Eyd. and Soul., t. 40, f. 6; t. 46, f. 5, 6.
caeruleus, Lea, t. 12, f. 3.
consobrinus,[7] Lea, t. 41, f. 7.
corbis, Bn. t. 45, f. 10.
corrugatus,[8] Müll. t. 44, f. 5, 6; t. 45, f. 2 to 5.
crispatus, Gould, t. 45, f. 1.
crispisulcatus, Bn. t. 11, t. 5.
exoloscens, Gould, t. 197, f. 5.
favidens,[9] Bn. t. 11, f. 1, 2, 3; t. 41, f. 3; t. 42, f. 2.
Feddeni,[10] Th. (River Penngunga. Cl. India.)
foliaceus, Gould, t. 42, f. 3.

---

[1] The Editors do not acknowledge the validity of many of these species, but merely illustrate them ; as to the arrangement it does not pretend to be scientific, but useful for grouping the allied forms of shells, not mollusks. Some species only just published have been added in our last part to fill up vacancies in the plates.

[2] Journ. Asi. Soc. Ben. 1873, pt. 2. pl. 17.

[3] In Prime's monograph of this genus we find recorded as Indian five of his species, which are wholly unknown to us ; C. subradiata (As. Lyc. N. York, 1861, vol. 8, p. 75, f. 23), C. Agnesis (ib. f. 23), C. parvula (ib. p. 76, f. 23), C. consanguinea (ib. 1867, vol. 8, p. 417), C. imperialis (ib. 1869, vol. 9). The last is stated to come from Pondicherry, always a suspicious locality ; the two first are probably immature, the two next insufficiently defined. Chemnitz (Conch. Cab. vol. 6, f. 324) has erroneously identified as the fluviatilis of Müller, a Tavoyer shell which may possibly be intended for occidens. The Tehrosian being estuary shells (C. Cyprinoides, Gray, and C. Cochinensis, Han.) are purposely omitted.

[4] Lea justly remarks that although edentulous when mature, the young have manifest teeth. The name has been changed

because D'Orbigny had previously published a Solenformis in Guérin's Mag. de Zool.

[5] Monosondylora Ava, Th. J. Asi. Soc. Beng. 1873, pt. 2, p. 209, pl. 17, f. 5.

[6] It is possible that the U. luteus of Lea (J. Ac. Philad. n. s. vol. 3, pl. 24, f. 4) from Newville, Tavoy, may be identical with one of our list, but the specimen delineated was in too bad a condition for positive recognition.

[7] The U. consobrinus of Kuster's Chemnitz (1 mo. p. 245, pl. 31, f. 2) has somewhat the aspect of this shell ; it is said to come from that indefinite locality the " E. Indies."

[8] Mörch states the U. gibbus of Spengler (Skriv. Nat. Kiobn. vol. 3, pl. 1) said to come from Tranquebar, is allied to this ; its latin description is utterly inadequate for the purpose of identification.

[9] To this may be referred the U. Mysolensis of Vanden Busch in Kuster's monograph of Unio (vol. Chenm. 1, pl. 78, f. 1). Perhaps, too, the U. Bagahensis of Lea (As. Un. vol. 3, p. 77, pl. 23, f. 534) which has been vainly sought for in the Rajah's Tank near Calcutta, its reputed locality, may prove, if indigenous, a distorted form of this or some allied species.

[10] J. Asi. Soc. Beng. 1873, pt. 2, p. 208, pl. 17, f. 3.

generosus, Gould, t. 46, f. 4, 7; t. 9, f. 7 (as lamellatus, var.) Gowhatensis,[1] Th. (R. Gowhattan, Assam). Isthens, Sow. t. 107, f. 1, 1. involutus, Bu. t. 41, f. 2. Jenkinsianus, Bu. t. 41, f. 4. lamellatus, Lea. t. 44, f. 7. Layardi, Lea, t. 41, f. 1. leinma,[2] Bu. t. 12, f. 6. mucilentus, Bu. t. 10, f. 2, 4; var. t. 154, f. 5. Matabayanus, Th. t. 154, f. 4. aureous, Ham. t. 42, f. 4, 5, 6.

marginalis,[3] Lam. t. 42, f. 7; t. 43, f. 2 to 5; t. 44, f. 1 to 4. Nuttallianus, Lea, t. 41, f. 5, 6. olivarius, Lea, t. 10, t. 1. pachysoma, Bu. t. 12, f. 1. parma, Bu. t. 154, f. 1; var.? t. 154, f. 2. pugio, Bn. t. 10, f. 7. raduka, Bu. t. 10, f. 3. rugosus, Gmelin, t. 154, f. 5. scobina, Bn. t. 46, f. 2. scutum, Bu. t. 46, f. 1; var. t. 46, f. 3.

Sikkimensis, Lea, t. 11, f. 4; t. 107, f. 6, 7. smaragdites, Bu. t. 10, f. 5. Tavoyensis, Gould, t. 154, f. 6, and var.? f. 7. Tennentii, Ham. t. 45, f. 7, 8, 9. theva, Bu. t. 12, f. 5. Thwaitesii, Bu. t. 43, f. 1. triembolus,[4] Bu. t. 107, f. 2. trinostris,[5] Ham. t. 11, f. 6. Valeanus, Ham. t. 155, f. 3. Wynegungensis, Lea, t. 45, f. 6.

INOPERCULATED LAND SHELLS.[6]

VITRINA.

Ataranensis,[7] Th. Bensoni, Pf. t. 65, f. 1, 4. Birmanica, Phil. t. 152, f. 7. cassida, Bu. t. 152, f. 2, 3. Ceylanica,[8] Beck. Christianae, Th. t. 66, f. 7, 10. Edgariana,[9] Bn. Flemingiana, Pf. t. 66, f. 5, 6. gigas, Bu. t. 66, f. 2, 3. heterosomcha, H. Bl. t. 152, f. 8, 9. arauitiana, Pf. t. 66, f. 8, 9. monticola, Bu. t. 152, f. 1, 4.

membranacea, Bu. t. 152, f. 10. ovata,[10] H. Bl. Pegurnsis,[11] Th. t. 65, f. 2, 3. prostans, Gould, t. 65, f. 5, 6. Salins, Bu. t. 65, f. 8, 9. scutella Bu. t. 66, f. 1, 4. solida,[12] G. A. t. 152, f. 6. succina, Reeve, t. 65, f. 7, 10. venusta, Th. t. 152, f. 5.

SORTINA.

Calus, Bu. t. 147, f. 2, 3, and var. discoidalis, 7. conjungens, Stolic. t. 145, f. 8, 9.

fenabilis, Bu. t. 147, f. 1, 4. schistostelis, Bu. t. 147, f. 5, 6.

HELIX.[13]

aralles, Pf. t. 128, f. 1, 4. achatina, Gray, t. 13, f. 1; t. 57, f. 8, 9. acris, Bu. t. 54, f. 6. acuminata, Bu. t. 50, f. 5. Akonteongensis, Th. t. 15, f. 4. albizonata, Dohrn, t. 52, f. 6. angulia, Bu. t. 25, f. 1. anax, Bu. t. 57, f. 1, 2, 3. auceps, Gould, t. 50, f. 1.

[1] J. Asi. Soc. Beng. 1873, pt. 2, p. 208, pl. 17, f. 4

[2] Compare with this and uusci natus the V. Shardhiana of Lea (J. As. Philad. n. s. vol. 5 (Obs. U. vol. 6) p. 23, pl. 27, f. 17), from Nagpore, and with this and caerulens his C. nyliatus (Obs, I. vol. 12, p. 59, f. 92).

[3] Müreh regards the V. trinostris of Spengler as a form of this most variable species.

[4] Compare with this the tripartitus of Lea (Proc. Philad. 1864), from the river Jellingbar.

[5] The trinostris of Reeve seems caerulens, to which species the semiplicatus of Troschel (Wiegm. Arch. Nat. 1837, p. 189); it a Unio at all, for the description remains one of Novaculina, may possibly belong.

[6] Linnea and Parmacella have been omitted for want of materials; according to Benson's manuscript the Vitrina bacenta of Hutton is the fry of P. rutella. Our work being to illustrate shells not mollusks, the rudimentary Helicarion Theobaldi of G. Austen (Proc. Z. 1872) and his H. crassus (517, pl. 30, f. 9) have been omitted. No 5 one are we for space.

[7] J. Asi. Soc. Beng. 1870, vol. 39, pt. 2, p. 401. From near the river Ataran, Marulaon.

[8] In Appendix to Index Moll. Chr. Frod. p. 1. Reeve

(Conch. Icon. Vit. t. 67) has figured what may possibly be the fry of a species described as twice or thrice the size of Ambanda. The plate referred to by Pfeiffer and Reeve has not been issued; indeed very few copies of even the Appendix were distributed.

[9] An. Nat. His. 1855, ser. 2, vol. 12, p. 93. Ceylon.

[10] J. Asi. Soc. Beng. 1871, vol. 40, pt. 2, p. 41, pl. 2, f. 7 (as Helicarion). Darjiling.

[11] Our figures precisely represent the type of the V. uniformis of Blanford (J. Asi. Soc. Beng. 1865, vol. 36, pt. 2, p. 26) from the Nilgherries.

[12] See, too, J. Asi. Beng. 1875, t. 2, f. 5. In that plate are delineated, also, Helicarion Shillinghense, brunneus and Nagurese, the shields of which hardly merit to be called shells.

[13] In addition to those in our list we omit specify the African H. Perotteti of Pfeiffer (Reeve and Kuster's figures are very unlike), and the Singapore H. lychnia of Benson, both ascribed to the Nilgherries; the Sanius fragilis of Hutton, which is most inadequately (J. Asi. Beng. vol. 7, p. 216) defined. Compare with our minute trochiform species the H. (Kaliella) policantha of Mörch (Attryk. Nat. Kjobenhavn. 1872) from Calcutta, which we have not seen.

[1] J. Asi. Soc. Beng. 1875. vol. 44. pt. 2, p. 2, pl. 1, f. 2.
[2] The edition of Blanford (J. Asi. Soc. Beng. 1865. vol. 34) is a synonym, teste Stoliczka.
[3] J. Asi. Soc. Beng. 1871. vol. 39, p. 352, pl. 1, f. 7; Cont. Mal. pt. 2. Pf. Mon. Hel. vol. 5, p. 85; Sowerby, and Kobmuller.
[4] Proc. Zool. 1857, p. 107. The types are identical.
[5] The type of Reeve's (C. Icon. Hel. f. 115) Shand of Sankeyet's Toornanensis most closely resembles our shell.
[6] Reeve's sculptured shell does not harmonize with the six striatula of Pfeiffer; erase the reference.
[7] Compare with this and its allies the H. Nagporensis of Pfeiffer (Pr. Z. 1860; Mon. H. vol. 5). The type (seen only as a distinct species) cannot be found.
[8] The Indian locality of H. albino is Pf. (Mon. Hel. vol. 5,

p. 381) was probably based upon the fact that one of his three types was the young of this or Reeve's, the other two come nearer the H. obtecta from Madeira.
[9] The H. revoluta of Pfeiffer (Mon. Hel. vol. 5, p. 416) is referred to this by Stoliczka.
[10] The two shells figured by Reeve as cysis and astsa (Pro. Z. P. 1851, p. 286; Reeve, f. 1382) are not even varieties; one own shell is his H. Anspaluroides (C. Ic. Helix, f. 1428), which many, and perhaps rightly, will think distinct.
[11] The variety fasciata of Godwin-Austen (J. Asi. Soc. Beng. 1875 (4), pt. 2), p. 3, pl. 1, f. 1, most surely be distinct.
[12] Godwin-Austen, Pr. Zool. 1875. p. 44.
[13] The Orobia Anhinamensis of Tryon (Amer. J. C. vol. 4, p. 139, p. 10, f. 4) is referred to this by Stoliczka.

¹ Compare with this and crinigera the Nilgherry H. triloba of Pfeiffer (Mon. Hel. vol. 4, p. 27; Novit. vol. 1, pl. 12, f. 11, 12). It is larger and not keeled.

² The H. Chamberti of Tryon (Amer. J. C. vol. 3, pl. 10, f. 2) is a synonym.

³ In Proc. Bost. 1843, p. 139, and Bost. J. Nat. H. vol. 4, p. 453, pl. 24, f. 6. P660. Mon. Hel. vol. 4. p. 452. Reeve, C. Icon. Hel. f. 770. The variety Tickellii only differs in being more carinated, and having the twin teeth close together.

⁴ The true H. Himalayana of Lea (Obs. 1 vol. vol. 1, pl. 19, f. 66) is referred by Hanley to this species; but Benson's erroneous ideal (Zool. J. vol. 5) was the H. cicatricosa.

⁵ J. As. Soc. Beng. 1872. vol. 41, pt. 2, p. 334, pl. 11, f. 9, 10; Kunnah Hills, Sunkoway, Arracan Hills.

⁶ See Pf. Mon. Hel. vol. 3, p. 73 and Nov. Ch. Hel. pl. 84, f. 23, 25.

⁷ If this should prove, as is possible, the true refuga of Gould, that name has priority; it resembles the refuga of Reeve. Philippi, and Kuster; yet the latter (Hel. t. 66, f. 21, 23) does not show the lamella correctly, nor the spiral sculpture, and almost equally with the second makes the outline of the last whorl opposite the mouth as prominent below as above, instead of sinuous inwards below as in leiophis.

⁸ The unique example from Trangitebar of the Nanina terrestris of Beck (Ap. Mol. Christ. Frod. p. 5, and Mörch, J. Con.

vol. 20 (1872, p. 344), said to remind one of this shell, being in the Danish Museum is inaccessible to us.

⁹ The H. eveys of Benson (An. Nat. 1859, ser. 2, vol. 5, p. 205) is the young of this shell.

¹⁰ The H. seposita of Benson (An. Nat. Hist. 1859, April) is only known to us by a sketch of Benson's, which precisely resembles shells given us by Blanford as his mucosa.

¹¹ Z. P. 1874, p. 610, pl. 73, f. 6, as M. (Pf-et.).

¹² Z. P. 1874, p. 609, pl. 73, f. 4, as M. (Pf-et.).

¹³ Benson stated that the Indlula of Reeve represents this shell.

¹⁴ In the Zoology of Belangee's Voyage we find (Moll. p. 414, pl. 1, f. 8, 9, 10) a H. senifissa, from near Pondicherry. Its figure reminds us of this shell, but it is said to have eight or nine whorls (four more than are represented). That of Pfeiffer is really Indian, and quite distinct. The H. inavelia of Benson (An. Nat. 1855, p. 93) is generally considered a form of partita, or some near ally (subspecies &c.). The date is subsequent to that of the species so named by Shuttleworth (1852).

¹⁵ Pr. Zool. 1869, p. 446. Near senis; from Bhotso.

¹⁶ An. Nat. H. 1853, p. 93. Pf. Mon. Hel. vol. 4, p. 59, found e'chan.

¹⁷ An. Nat. H. 1859, p. 390. Pf. Mon. Hel. vol. 5, p. 69; from Phie Than, Tenasserim.

¹⁸ Benson erroneously identified this shell with the imperforated five-whorled H. vitrinoides of Deshayes (Mag. de Z. 1831, pl. 26), of which the locality was then unknown. In Kuster's

Phoenix, Pf. t. 127, f. 6.
phyllophila, Bu. t. 61, f. 10.
palidiua, Bu. t. 53, f. 6.
pinacis,[1] Bu. t. 13, f. 5; t. 84, f. 1, 1.
Pirricana, Pf. t. 87, f. 5, 6.
planiscula, Hut. t. 32, f. 7, 10.
plectostoma, Bu. t. 13, f. 2.
plicatula, Bl. t. 28, f. 1.
politissima, Bu. t. 31, f. 8, 9.
Pollux, Th. t. 26, f. 2, 5.
polypleuris, Bl. t. 16, f. 7.
Poongee, Th. t. 16, f. 9.
propinqua, Pf. t. 30, f. 10.
prospera, Albers, t. 150, f. 4.
proxima, Fér. t. 28, f. 5.
pseudophis,[3] Bl. & G. A.
pulchella, Müll. (Cashmire).
pylaica, Bu. t. 15, f. 2.
radicicola, Bu. t. 62, f. 10.
refuga,[3] Gould, (Tavoy, Birmah).
regulata, Bu. t. 31, f. 5, 6.
repercussa, Gould, t. 13, f. 4.
resplendens,[4] Phil. t. 51, f. 4.
retifera, Pf. t. 87, f. 8, 9.
retrorsa, Gould, t. 25, f. 6.
rivicola,[5] Bu. t. 61, f. 1.
Rivolii, Desh. t. 14, f. 2.
rorida, Bu.[6]
Rosamundu, Bu. t. 59, f. 5, 6.
rotatoria, V. Busch, t. 15, f. 5.
rubellocincta, Bl. t. 51, f. 5, 6.
ruginosa, Fér. t. 84, f. 2, 3.
sanis, Bu. t. 83, f. 1, 7.
Saturnia, Gould, t. 25, f. 3.
sculpturita, Th. t. 53, f. 9.
scenosma, Bu. t. 53, f. 3, 4, 5.
semidecussata, Pf. t. 58, f. 1, 2.

semirugosa, Beck, t. 59, f. 4.
sequax, Bu. t. 63, f. 1, 2, 3.
sericata,[7] G. A. t. 132, f. 8, 9.
serrula, Bu. t. 59, f. 7.
Shanensis, Stol. t. 149, f. 8, 9.
Shiplayi, Pf. t. 131, f. 7, 10.
Sheroiensis, G. A. t. 159, f. 7.
Shisha,[8] G. A. (Naga and Khasi
      Hills).
similaris,[9] Fér. t. 53, f. 1, 2.
Sisparica, Bl. t. 112, f. 4, 5, 6.
Skinneri, Reeve, t. 141, f. 1.
solata, Bu. t. 28, f. 6.
splendens, Hutt. t. 51, f. 7, 10.
stephus, Bu. t. 62, f. 4, 5, 6.
subcostoidea, Pf. t. 84, f. 7, 10.
subcornea, Pf. t. 149, f. 2, 3.
subdecussata, Pf. t. 56, f. 4.
subjecta, Bu. t. 64, f. 1, 2, 3.
superba, Pf. t. 127, f. 4.
tapeina, Bu. t. 15, f. 5.
Taprobanensis, Dohrn, t. 29 f. 2.
tenuicula, Adams, t. 89, f. 7, 10.
tertiana, Bl. t. 16, f. 10.
textrina, Bu. t. 52, f. 2, 3.
Theodori, Phil. t. 59, f. 7, 8.
Thwaitesii, Pf. t. 128, f. 7, 10.
thyreus, Bu. t. 27, f. 6.
todarum, Bl. t. 64, f. 4, 5.
Tranquebarica, Fab. t. 59, f. 3.
Tavanorensis, Bu. t. 50, f. 5, 6; t.
      149, f. 7.
tricarinata, Bls. t. 129, f. 7, 10.
trifasciata, Bl. t. 131, f. 2.
trifilosa,[10] Pf.
trilamellaris,[11] G. A.
trochalia,[12] Bu. t. 28, f. 7.

turritella, Ad. t. 86, f. 4.
tugurium, Bu. t. 29, f. 10.
umbrina, Pf. t. 89, f. 1, 2, 3.
umbosa, Bl. t. 111, f. 2, 3 (var.).
uter, Th. t. 58, f. 7, 8.
vallicola, Pf. t. 128, f. 8, 9.
vermeula, Pf. t. 150, f. 9.
vesicula, Bu. t. 62, f. 4, 5, 6.
vidua, Bl. t. 130, f. 2, 3.
vilipensa, Bu. t. 89, f. 4, 5, 6.
vitellinus, Pf. t. 59, f. 1, 2.
vittata, Müll. t. 130, f. 10.
Waltoni, Reeve, t. 127, f. 1.
Woodiana, Pf. t. 90, f. 2, 3.
Zoroaster, Th. t. 86, f. 2, 3.

BOYSIA.

Bensoni, Pf. t. 8, f. 1.

HYPSELOSTOMA.

Bensonianum, Bl. t. 8, f. 2.
Dayanum, Stol. t. 147, f. 10.
tubiferum, Bu. t. 8, f. 3.

STREPTAXIS.

Andamanica, Bu. t. 8, f. 6.
Blanfordi, Th. t. 8, f. 5, as Birmanica.
boudleax, Bu. t. 156, f. 9; t. 31, f.
      1, 4 (young).
Burmanica, Bl. t. 156, f. 10; var.
      edentula, t. 8, f. 10, as Blanfordi.
Canarica, Bed. t. 156, f. 7.
Cingalensis, Bu. t. 98, f. 2, 3.
exarata, Gould, t. 98, f. 8, 9.
Elisa,[13] Gould.
Layardiana, Bu. t. 98, f. 1, 4.

Chennitis (Helix, pl. 110) the true vitrinoides (t. 13. f. 14. 15).
and the spurious (f. 10, 11, 12) are both delineated as one species.

[1] The H. (Corilla) pettos of Martens (Mal. Bla. 1868, p. 138) from the Himalaya may be compared with this and its allies.

[2] Z. S. 1874, p. 610, pl. 74, f. 3, as H. (Fleet.) Thayet. Myo, Pegu. Compare H. refuga of Philippi, &c.

[3] Proc. Bost. vol. 2, 1846, p. 99; Otia, p. 198. The species (one or more) figured by Philippi (Ab. Conch.), Reeve, and Kuster (Hel. t. 66, f. 21, 22, 23) does not suit the "similicornea" of our author, who states it is almost exactly like "carabiata" (Rivolii). Our supposed dextral var. has been lately named dextrorsa.

[4] This is the H. expedita of Deshayes (Fér. H. Moll. pl. 87, f. 4). The resplendens of Beck, said to come from Bombay, was described, but not strictly published in the Appendix to his Catalogue. Compare our figure 2 of pl. 51 with his definition.

[5] The H. (Ariesta) elusior of Martens (Mal. Blat. 1868, p. 137) is, probably, this shell in fine condition.

[6] Ac. Nat. 1859, v. 3, vol. 3, p. 266. Pf. f. Mon. Hel. vol. 5, p. 444; Bayding, and Sandal Hill.

[7] Pr. Z. 1871, p. 608, pl. 73, f. 5.

[8] J. Asi. Soc. Beng. 1875, vol. 44, pt. 2, p. 2, pl. 1, f. 3.

[9] To this may be referred the H. smilsis of Pfeiffer chon.
vol. Chem. Hel. pl. 60, f. 19, 20.

[10] Proc. Zool. 1853, p. 125; Mon. Hel. vol. 4, p. 47; Ceylon.
Cleotype could not be found; our shell would be retifera with a narrower perforation.

[11] Proc. Zool. 1876, p. 15.

[12] The Anguilla Bigsbyi of Tryon in the Amer. J. C. vol. 5, p. 110, pl. 10, f. 3.

[13] Proc. Bost. 1856, p. 12; Otia, p. 220; from an islet in the Mergui Archipelago.

[1] J. As. Soc. Beng. 1874, vol. 40, pt. 2, p. 166, pl. 7, f. 11, 12, 17.

[4] Compare the S. Festiva, or Perotteti Festiva var. of the Blanford's (J. As. Soc. Beng. 1861, p. 558, pl. 2, f. 6) from Southern India.

[3] Our supposed variety (pl. 8, f. 7) has been separated by Stolizcka as S. Hanleyana (J. As. Beng. 1874, pl. 7, f. 15); he is quite right in regarding the form with a more tapering and rounded mouth as better suited to Benson's description (Pf. if. Mon. Hel. vol. 3, p. 112), although Benson's drawing (in our possession) represents the other. Had we then possessed the 'typical Sankeyi, we should have preferentially figured it; it is very like our pl. 98, f. 10, only has not a smooth base or a sexual base-like calcycle. Our figured specimen is quite as large as the more characteristic form (J. As. Beng. 1874, vol. 40, pt. 2, p. 167, pl. 7, f. 14).

[4] Bens. Am. Nat. H. 1849, ser. 2, vol. 4.—Pfeiff. Mon. Hel. vol. 3, p. 560; Kust. ed. Chemn. Pup. pl. 17, f. 21, 22. Burrakpur, Bengal.

[5] Am. Nat. Hist 1863, ser. 3, vol. 11, p. 427.—14 if. Mon. Hel. vol. 6, p. 319. Central India.

[2] Identical with the undescribed ignota of Th.
J. As. Soc. Beng. 1872, vol. 41, pt. 2, p. 201, pl. 9, f. 5. Toughu ; Birmah.

[8] Proc. Bost. N. H. 1859, vol. 6, p. 12 : Otia, p. 220. Pt. Mon. vol. 6, p. 409; from Tavoy; has six whorls only. Can it be Philippiana? Compare also the vespa of J. As. Soc. Beng. 1872, vol. 41, pt. 2, p. 200, pl. 9, f. 15.

[7] J. As. Soc. Beng. 1872, vol. 41, pt. 2, pl. 9, f. 19. Changhighalli, near Maré, W. Himalayah.

[10] See also, the B. Huttoni, Pf. (Symb. pt. 3, p. 54 : Mon. Helix vol. 2, p. 148 : Kust. ed. Chz. Bul. t. 17, f. 3, 4) from the "E. Indies."

[7] To this group belongs the B. Mahabanicus of Pfeiffer (Mal. Blät. 1857, p. 159); Mon. Hel. vol. 4, p. 414, from Ahmednuggur.

[12] Proc. Bost. 1843, vol. 1 (copied Otia, p. 189) : Bost. J. Nat. H. vol. 4, p. 457, pl. 24, f. 3.—Reeb. Nr. Mol. t. 139, f. 9, 10.

[13] The types of the B. barbus of Pfeiffer (Proc. Z. 1852, p. 157), from the "E. Indies," belong to this species.

Janus,[1] Pf. t. 19, f. 5 (as atriculkosus).
Jerdoni,[2] Bu. t. 21, f. 7.
Khasianus, G. A. t. 148, f. 7.
Kumawarensis, Hutt. t. 19, f. 3.
latebricola, Bu. t. 79, f. 7.
Layardi, Bu. t. 79, f. 2, 3.
lepidus, Gould, t. 80, f. 6.
lubricus, Brug. (Cashmire).
Mavortius, Reeve, t. 148, f. 5.
Moussonianus, Pf. t. 21, f. 4.
Munipurensis, G. A. t. 148, f. 1, 4.
Nilagiricus, Pf. t. 23, f. 3.
nivicola, Bu. t. 22, f. 9.
orbus, Bl. t. 20, f. 1.
Pertica, Bu. t. 22, f. 7.
physalis, Bu. t. 21, f. 9.
plicifer, Bls. t. 80, f. 8.
praetermissus, Bls. t. 19, f. 4.
pretiosus, Cant. t. 23, f. 7.
proletarius,[3] Bu. t. 80, f. 3.
punctatus,[4] Ant. t. 20, f. 10.
pusillus,[5] Bl. t. 79, f. 8.
putus, Bu. t. 80, f. 9.
rufopictus, Bu. t. 21, f. 10.
rufostrigatus, Bu. t. 20, f. 4; t. 23, f. 10.
salsicola, Bu. t. 20, f. 8.
scrobiculatus, Bl. t. 79, f. 9.
segregatus, Bu. t. 80, f. 10.
Sikkimensis, Bu. t. 19, f. 7.
Sindicus, Bu. t. 20, f. 6.
Simensis,[6] Bu. t. 21, f. 5, 6.
Smithei, Bu. t. 20, f. 3.
spelaeus, Hutt. t. 23, f. 8.
stalix, Bu. t. 22, f. 3.

Syllaeticus, Reeve, t. 19, f. 9.
tetrebralis,[7] Th.
Theobaldianus,[8] Bu. t. 19, f. 10.
trifasciatus, Beng. t. 21, f. 3.
tralila, Bl. t. 80, f. 4.
vibex, Hut. & Bu. t. 22, f. 8; t. 23, f. 3; t. 20, f. 5.
vicarius, Bl. t. 22, f. 2.
Walkeri, Bu. t. 79, f. 4.

CORILLSTRE.

scalaris, Bu. t. 156, f. 5.

ACHATINA.

ancentum, Bu. t. 35, f. 3.
Anamullica,[9] Bl.
Arthurii, Bu. t. 36, f. 3.
loculina, Bl. t. 78, f. 6.
balanus, Bu. t. 102, f. 10.
Beddomei, Bl. t. 102, f. 8; t. 156, f. 4.
Bensoniana, Pf. t. 102, f. 3.
butellus, Bl. t. 35, f. 4.
Bottampotana, Bed. t. 156, f. 1.
brevis, Pf. t. 18, f. 10.
Burrailensis,[10] G. A. (East Burrail Range).
Butleri,[11] G. A. (East Burrail Range.)
capillacea, Pf. t. 156, f. 3.
Cassiaca, Bu. t. 36, f. 5.
Ceylanica, Pf. t. 17, f. 4.
Chessoni, Bu. t. 18, f. 8.
corrosula, Pf. t. 18, f. 2.
crassilabris, Bu. t. 36, f. 1.
crassula, Bu. t. 36, f. 4.
Deshayesiana, Pf. t. 102, f. 2.
crosa, Bl. t. 78, f. 5.

faceula, Bu. t. 35, f. 1.
Fairbanki, Bu. t. 18, f. 3.
filosa, Bl. t. 35, f. 10.
fusca, Ad. t. 78, f. 4.
gemma, Bu. t. 36, f. 7.
hastula, Bu. t. 18, f. 1.
hebes, Bl. t. 156, f. 2.
Hugeli, Pf. t. 78, f. 2.
illustris,[12] G. A. t. 102, f. 9.
inornata, Pf. t. 17, f. 2, 3?
Isis, Han. t. 156, f. 5.
Jerdoni, Bu. t. 78, f. 10.
leptospira, Bu. t. 35, f. 2, 3.
lyrata,[13] Bl. t. 18, f. 9.
Mullorum, Bl. t. 102, f. 5.
nitens, Gray. t. 17, f. 1.
notigena, Bu. t. 35, f. 8, 9.
obtusa, Bl. t. 36, f. 6.
Oreas, Bu. t. 7, f. 9.
Orobia, Bu. t. 18, f. 7, 8.
orthoceras, G. A. t. 156, f. 6.
pachycheila,[14] Bu.
panacha, Bu. t. 36, f. 2.
parabilis, Bu. t. 35, f. 7.
pampereula, Bls. t. 102, f. 1.
Peguensis, Bl. t. 102, f. 6.
Perotteti, Pf. t. 35, f. 6.
pertenuis, Bl. t. 18, f. 5.
pinelustris, Bu. t. 17, f. 6, 7.
pulla, Bl. t. 78, f. 1.
Punctogallana, Pf. t. 102, f. 4.
pyramis, Bu. t. 18, f. 6.
rugata, Bl. t. 102, f. 7.
Sarissa, Bu. t. 35, f. 10.
serutillus, Bu. t. 18, f. 1.
senator, Han. t. 156, f. 4.

---

[1] Proc. Z. 1852, p. 85, with the misleading locality of New Hebrides. The type has been examined.

[2] Pfeiffer has referred his B. Redfieldi (Mal. Blät. 1854) to this species.

[3] A dwarf form of this precisely agrees with the B. Panos of Benson (Ann. N. H. 1853, ser. 2, vol. 2, p. 94.—Pf. Mon. II. vol. 4, p. 417).

[4] This B. acutus? of Hutton, from Jeypur, was probably this shell.

[5] The unique type of the Bulimulus pusillus of Adams (Pr. Z. 1867, p. 2, pl. 19, f. 17) proves to be identical with the British Balea perversa, which can scarcely be a native of Ceylon.

[6] The supposed B. citrinus so briefly referred to by Benson in the Asiatic Journal (Beng. vol. 5) was probably this, flavus, or Syllaeticus. The yellow form comes from Akyab, the banded from near Prome.

[7] J. As. Soc. Beng. 1870, vol. 39, pt. 2, p. 101. Sisur Provinces. Belongs to the Streptaxis section (Spraxis).

[8] Although the type does not at all agree with Gould's description of the mouth of his B. monilifers, it is certain that he sent a specimen of it so named to one of his correspondents. If identical, Gould's name must take priority.

[9] J. As. Soc. Beng. 1866, p. 27; Conch. Ind. pl. 6. Pf. vol. 6, p. 223. Anamullay Hills. Only a broken specimen known to us.

[10] J. As. Soc. Beng. 1874, vol. 44, pt. 2, p. 3, t. 1, f. 6.

[11] Id. p. 4, t. 1, f. 7.

[12] Described by G. A. in J. As. Soc. Beng. 1875, vol. 44, pt. 2, p. 5.

[13] See, also, J. As. Soc. Beng. 1870, vol. 39, pt. 2, p. 20 var. Gleeniana, pl. 3, f. 19, and var. Malleciana, from near Bombay.

[14] An. Nat. Hist. 1853, p. 94. Pf. Mon. II. vol. 3, p. 688, Ceylon.

serena, Bu. t. 78, f. 8.
Shiplayi, Pf. t. 96, f. 9.
Singhurensis, Bl. t. 78, f. 7.
subfusiformis.[1] Bl.

Tanuilica, Bl. t. 17, f. 9.
tennispira, Bu. t. 36, f. 8.
textilis, Bl. t. 17, f. 10.
Theobaldi,[2] Hanl. t. 17, f. 5.

Torrensis, Bl. t. 78, f. 3.
Vadalica, Bu. t. 35, f. 5.
vernina. Bn.[3]

OPERCULATED LAND SHELLS.

DIPLOMMATINA.

affinis. Th.[4]
Anamallayana,[5] Bedd.(Anamallays).
angulata, Th. & St. t. 140, f. 7 (enlarged).
Austeni, Bl. t. 119, f. 1, 4.
Blanfordiana. Bn. t. 119, f. 5, 6.
Burrii.[6] G. A. (Assam).
Canarica.[7] Beddome (N. Canara).
carneola, Stol. t. 140, f. 3 (enlarged).
Ceylanica,[8] Bedd. (Pedrotalle Galle, Ceylon).
convoluta.[9] G. A. (Eastern Burrail).
costulata, Hutt. t. 120, f. 8, 9.
crispata, Stol. t. 141, f. 6.
depressa, G. A. t. 120, f. 5, 6.
diplochcilus, Bu. t. 140, f. 2, 3 (enlarged).
exilis, Bl. t. 119, f. 10.
folliculus. Pf. t. 140, f. 8, 9 (enlarged).
Fairbanki, Bl. t. 141, f. 9.
gibbosa, Bl. t. 120, f. 1, 4.
gracilis,[10] Bedd. (Gudaon Hills,Vizagapatam).
Huttoni, Pf. t. 139, f. 5, 6 (enlarged).
insignis, G. A. t. 139, f. 10 (enlarged).
Jaintiaca, G. A. t. 120, f. 2, 3.
Jatingana,G. A. t. 139, f. 7(enlarged).
Kinginna, Bls. t. 141, f. 1.
labiosa, Bl. t. 139, f. 9.

liricincta, Bl. t. 141, f. 2.
minima,[11] Bedd. (Gudaon Hills, Vizagapatam).
nana, Bl. t. 140, f. 1 (enlarged).
Nilgirica. Bl. t. 141, f. 8.
nitidula, Bl. t. 141, f. 5.
oligopleuris, Bl. t. 119, f. 2, 3.
pachycheila, Bu. t. 140, f. 5, 6 (enlarged).
parcula, G. A. t. 139, f. 4 (enlarged).
Pedronis,[12] Bedd.
polypleuris, Bu. t. 140, f. 10 (enlarged).
pallula, Bu. t. 119, f. 7.
Pulneyana, Bl. t. 141, f. 3.
Paperformis,[13] Th.
Puppensis, Bl. t. 139, f. 8, 9 (enlarged).
Richthofeni, Stol. t. 141, f. 7, 8.
Salwiniana,[14] Th.
scalaris,Bl. t. 139,f. 2,3 (enlarged).
scalaroides, Th. t. 141, f. 10.
semisculpta, Bl. t. 120, f. 7.
Sherfaiensis, G. A. t. 119, f. 8.
sperata,[15] Bl.
subovata,[16] Bedd.
tumida, G. A. t. 139, f. 1 (enlarged).
ungulata, H. Bl. t. 120, f. 10.

STREPTAULUS.

Blanfordi, Bu. t. 133, f. 5, 6.

RAPHAULUS.

chrysalis, Pf. t. 133, f. 7.
pachysiphon, Th. & St. t. 133, f. 4.

CATAULUS.

aureus, Pf. t. 106, f. 9.
Austenianus, Bu. t. 106, f. 7.
Blanfordi, Dohrn, t. 106, f. 3.
Calendoensis, Bl. t. 106, f. 10.
decorus, Bu. t. 106, f. 5.
duplicatus,[17] Pf. t. 106, f. 2.
eurytrema, Pf. t. 146, f. 3.
haemastoma, Pf. t. 106, f. 4.
Layardi, Gray, t. 106, f. 8.
leucocheilus, Ad. and R. t. 146, f. 1.
marginatus, Pf. t. 145, f. 6.
pyramidatus, Pf. t. 146, f. 5.
recurvatus, Pf. t. 146, f. 2.
Templemani, Pf. t. 106, f. 1.
Thwaitesii,[18] Pf. t. 106, f. 6.

PUPINA.

artata, Bn. t. 7, f. 5.
aruda, Bn. t. 7, f. 4.
Blanfordi, Th. t. 7, f. 6.
imbricifera, Bn. t. 7, f. 7.
Peguensis,[19] Bn.

MEGALOMASTOMA.

funicolatum, Bu. t. 7, f. 2 ; t. 133, f. 1.
gravidum,[20] Bn. t. 7, f. 1.

[1] Pr. Zool. 1868, p. 449. = A. (Glessula). Parsee in Yunan.
[2] From Teria Ghat.
[3] An. Nat. Hist. 1843, p. 91.—Pf. Mon. Hel. vol. 4, p. 645; Nahuds, Ceylon.
[4] J. Asi. Soc. Beng. 1870, vol. 39, pt. 2, p. 398. Shan Provinces.
[5] Proc. Zool. Soc. 1875, p. 443, pl. 52, f. 5, 6.
[6] J. Asi. Soc. Beng. 1875, vol. 44, pt. 2, p. 8, pl. 4, f. 1.
[7] Proc. Zool. Soc. 1875, p. 442, pl. 52, f. 1.
[8] Id. 1875, p. 444, pl. 52, f. 9.
[9] J. Asi. Soc. Beng. 1875, vol. 44, pt. 2, p. 8, pl. 4, f. 8.
[10] Proc. Zool. 1875, p. 442, pl. 52, f. 2.
[11] Id. 1875, p. 442, pl. 52, f. 3, 4.

[12] Proc. Zool. Soc. 1875, p. 443, pl. 52, f. 8.
[13] J. Asi. Soc. Beng. 1870, vol. 39, pt. 2, p. 398. Shan Provinces.
[14] Id. 1870, vol. 39, pt. 2, p. 398. Shan Provinces.
[15] Id. 1862, vol. 30, p. 142 : Cat. Mal. pt. 3.— Pf. Mon. Pa. vol. 3, p. 10. Arakan Hills. Unique.
[16] Proc. Zool. Soc. 1875, p. 443, pl. 52, f. 7.
[17] Runs into Templemani.
[18] In this we include the C. Cmaingii of Pfeiffer (Proc. Z. el. 1856, p. 359.—Sow. Th. vol. 3, pl. 264, f. 3).
[19] Am. Nat. Hist. 1860, ser. 3, vol. 6, p. 192.- Pfeif. Mon. Pneum. vol. 3, p. 35; Pegu. In Calcutta Museum.
[20] The Operena Mounts of Benson is the young.

paupervulum, Bn. t. 133, f. 3.
sectilabrum, Gould, t. 7, f. 3.

POMATIAS.

Himalayæ, Bn. t. 7, f. 9.
Pegnensis, Th. t. 7, f. 8.
pleurophora, Bn. t. 7, f. 10.

CREMNOCONCHUS.

carinatus, Lay, t. 146, f. 10.
conicus, Bl. t. 146, f. 8.
Fairbanki, Bl. t. 146, f. 7.
Syhadrensis, Bl. t. 146, f. 6.

OMPHALOTROPIS.[1]
distermina, Bn. t. 145, f. 10.

ACMELLA.

hyalina,[2] Th.& St. (near Moulmein).
tersa, Bn. t. 117, f. 1.

HYDROCENA.[3]

Blanfordiana, Stol. t. 117, f. 2.
fraterna,[4] Th. & St.
frustillum, Bn. t. 117, f. 5.
illex, Bn. t. 117, f. 4.
liratula,[5] Stol.
pyxis, Bn. t. 117, f. 3.
Rawesiana, Bn. t. 117, f. 6.
saritta, Bn. t. 117, f. 7.

JERDONIA.[6]

Playrei,[7] Th. t. 135, f. 3.
trochlea, Bn. t. 135, f. 5, 6.

CRASPEDOTROPIS.

cuspidatus, Bn. t. 135, f. 1, 4.
fimbriata,* G. A. (W. Naga Hills).
Salemensis,[8] Bedd. (Salem district,
S. India).

MYCHOPOMA.

hirsutum, Bedd. t. 136, f. 1, 4.
fimbiferum, Bl. t. 136, f. 2, 3.
seticinctum,[10] Bedd. (Anamallay
Mountains.)

CYATHOPOMA.

Ceylanicum, Bedd. t. 145, f. 8.
Coonoorense,[11] Bl. t. 135, f. 10.
Deccanense, Bl. t. 82, f. 8, 9.
filocinctum, Bn. t. 82, f. 2, 3.
Kairyenense,[12] W. and H. Blan.
Kolamullicaar, Bls. t. 135, f. 8, 9.
latilabre,[13] Bedd.(S.Canara Ghats).
Malabaricum, Bls. t. 82, f. 1, 4.
malleatum, Bls. t. 82, f. 5, 6.
procerum, Bl. t. 135, f. 7.
Shevaroyanum,[14] Bedd. (Salem dis-
trict, S. India).

tiguarium, Bn. t. 82. f. 7, 10.
Travancoricum,[15] Bedd.(Travancor-
Mountains).
vitreum, Bed. t. 145, f. 9.
Wynaadense,[16] Bl.

OPISTHOSTOMA.

Fairbanki, Bl. t. 117, f. 8.
macrostoma. Bed. t. 117, f. 9.
Nilghiricum,[17] Bls. t. 117, f. 10.

ALYCÆUS.[18]

amphora, Bn. t. 91, f. 2, 3.
Andamanim, Bn. t. 91, f. 7, 10.
armillatus, Bn. t. 93, f. 10.
Avæ, Bl. t. 94, f. 8, 9, 10.
bembex, Bn. t. 95, f. 2, 3.
bicrenatus,[19] G. A.
bifrons, Th. t. 93, f. 1, 4.
Burtii,[20] G. A.
conicus, G. A. t. 103, f. 8, 9.
constrictus, Bn. t. 95, f. 1, 4.
cronatus, G. A. t. 103, f. 2, 3.
crenulatus, Bn. t. 97, f. 1, 4.
crispatus,[21] G. A.
cucullatus, Th. t. 96, f. 1, 4.
diagonus, G. A. t. 103, f. 1.
digitatus,[22] Bl.

---

[1] The O. zonatus of Desh. (Voy. Bélang. Z. 416, pl. 1, f.
16, 17, as Cyclost.) is a native of the Isle of Bourbon, and not
from "Pondicherry."

[2] J. Asi. Soc. Beng. 1872, vol. 41, pt. 2, p. 333, pl. 11, f. 7.

[3] The type of the H. unifasa of Benson (An. Nat. 31, 1853,
ser. 2, vol. 11, p. 158), figured by Godwin-Austen (Proc. Zool.
1872, pl. 30, f.3) is the fry of some large shell, possibly of a Pupa.

[4] J. Asi. Soc. Beng. 1875, vol. 44, pt. 2, p. 332, pl. 11, f. 5,
6. At san Valley, near Moulmein.

[5] Id. 1871, vol. 40, pt. 2, p. 157, pl. 6, f. 5. Pamotha, near
Moulmein.

[6] At the last moment, and too late for redistribution into our
groups, a paper (Proc. Zool. 1879) on Cyathopoma (in a broader
sense than usual) has reached us from Col. Beddome. The fol-
lowing species are regarded by him as belonging to Jerdonia,
which he regards as a section of the genus: nitidum, Blanfordi,
album (? our Kolamullicaar), Anamallayanum, ovatum, Svag-
gherianum, microstoma, clavus ; they are all figured in pl. 56,
and if correctly delineated, such shapes as 14 and 17 can scarcely
rank as members of the same genus.

[7] The Cyclop. billatus of Beddome (Z. P. 1875, p. 452, pl.
53, f. 34 is a synonym.

[8] J. Asi. Soc. Beng. 1875, vol 14. pt. 2. p. 7, pl. 4, 1.

[9] Proc. Zool. 1875, p. 453, pl. 53, f. 35 (as Cyclostoma)

[10] Id. 1875. p. 449, pl. 53, f. 23. 24 (as Cyathop.)

[11] The striæ should be closer, and not visible at the mouth
the ridges should be more marked and fewer.

[12] J. Asi. Soc. Beng. 1861, vol. 29, p. 352, pl. 2, 1. 4 and
Cent. Mal. pt. 2, as Cyclost. — Pfeif. Mon. Pn. vol 2. p. 55, as
Cyclost.— Bl. J. Conch. 1868, pl. 12, f. 4, as Cya. Kairyen. Bed.
near Sabsa, S. India.

[13] Proc. Zool. 1875, p. 450, pl. 53, f. 28, 29.

[14] Id. 1875, p. 451, pl. 53, f. 32, 33.

[15] Id. 1875, p. 451, pl. 53, f. 30, 31.

[16] Jour. Conch. 1868, p. 259, pl. 12, f. 3. Wynaad N....
of Nilgherries.

[17] Compare O. Deccanense, Bedd. Proc. Zool. 1875. p. 444
pl. 52, f. 10, 11. See, too, his O. distortum, p. 445.

[18] In the J. Asi. Soc. Beng. 1871, vol. 40, pt. 2, will be found
drawings of A. Ingensi, var. Nagaensis (pl. 4, f. 5) and A.
otiphorus (pl. 5, f. 6).

[19] Id. 1874, vol. 43, pt. 2, p. 148, pl. 2, f. 5.

[20] J. Asi. Soc. Beng. 1874, vol. 43, pt. 2, p. 149, pl. 3, f. 9

[21] Id. 1871, vol. 40, pt. 2, p. 91, pl. 4, f. 1, and f. 2 (var.)
Khasia Jaintia and N. Cachar Hills.

[22] Id. 1871, vol. 40, pt. 2, p. 41, pl. 2, f. 4. Imphling.

b

expatriatus, Bl. t. 145, f. 1, 4.
Feddenianus, Tu. t. 91. f. 1, 4.
Footei,[2] Bls.
gemmula, Bn. t. 93, f. 7.
glaber, Bl. t. 97, f. 8, 9, 10.
globulus,[3] G. A.
graphicus, Bl. t. 96, f. 7, 8, 9.
hebes, Bn. t. 93, f. 5, 6.
humilis, Bl. t. 93, f. 8, 9.
Jaintiacus,[3] G. A.
inflatus,[4] G. A.
Ingrami, Bl. t. 92, f. 7, 10.
Khasineus, G. A. t. 103, f. 5, 6.
Kurzianus, Th. & St. t. 145, f. 2, 3.
margarita, Th. t. 95, f. 10: t. 97, f. 7.
multirugosus,[5] G. A.
nitidus, Bl. t. 94, f. 4, 7.
otiphorus, Bn. t. 96, f. 5, 6.
physis, Bn. t. 92, f. 5, 6.
plectocheilus, Bn. t. 96, f. 5, 6.
politus, Bl. t. 94, f. 1, 2, 3.
polygonoma, Bl. t. 96, f. 2, 3.
prorectus, Bn. t. 92, f. 2, 3.
pusillus, G. A. t. 103, f. 7, 10.
pyramidalis, Bn. t. 94, f. 5, 6.
Richthofeni, Bl. t. 94, f. 5, 6.
serratus,[6] G. A.
sculptilis, Bn. t. 97, f. 5, 6.
Stoliczkii,[7] G. A.
strangulatus, Hutt. t. 93, f. 2, 3.
strigatus,[8] G. A.
sculpturus,[9] G. A. (Manipur).

succineus, Bl. t. 96, f. 7, 10.
stylifer, Bn. t. 92, f. 1, 4.
Theobaldi, Bl. t. 97, f. 2, 3.
umbonalis, Bn. t. 92, f. 8, 9.
urnula, Bn. t. 91, f. 8, 9.
vestitus, Bl. t. 103, f. 4.
Vulcani, Bl. t. 96, f. 8, 9.

PTEROCYCLOS.[10]

Andersoni (Spir.) Bl. t. 49, f. 3, 4.
ater, St. t. 142. f. 5, 6.
Avanus, (Spir.), Bl. t. 134, f. 8, 9.
Beddomei (Spir.), Bl. t. 134, f. 5, 6.
léfrens, Pf. t. 142, f. 8, 9.
bilabiatus, Sow. t. 5, f. 2.
ectra, Bn. t. 134, f. 7, 10.
Cingalensis, Bn. t. 5, f. 5.
Cumingii, Pf. t. 49, f. 7, 8.
Fairbanki, (Spir.), Bl. t. 49, f. 1, 2.
Feddeni, Bl. t. 5, f. 9; t. 134, f. 1.
Gordoni (Spir.), Bn. t. 49, f. 9, 10.
Haughtoni, (Rhies.), Bn. t. 5, f. 10.
hispidus, (Spir.) Pears. t. 5, f. 4.
insignis, Th. t. 5, f. 6, 7.
Massræi, Bl. t. 5, f. 1.
nanus, Bn. t. 49, f. 5, 6.
parvus,[11] Pears. t. 5, f. 3; t. 142, f. 5, 6.
pallatus, Bn. t. 134, f. 2, 3, 4.
rupestris, Bn. t. 5, f. 8.

AULOPOMA.

grande, Pf. t. 47, f. 1, 2.

Helicinum, Pf. t. 4, f. 8.
Hoffmeisteri, Guér. t. 47, f. 3, 4.
Itieri,[12] Tros. t. 4, f. 6, 7.
sphæroideum,[13] Dohrn (Ceylon).

LAGOCHILUS.

leporinus, Bl. t. 135, f. 2.
scissumargo, Bn. t. 6, f. 7.
tomotrema, Bn. t. 6, f. 8.

LEPTOPOMA.[14]

apicatum, Bn. t. 142, f. 1.
aspirans,[15] Bn. t. 6, f. 4.
comulus, Pf. t. 105, f. 1.
cybeus, Bn. t. 6, f. 1.
elatum, Pf. t. 142, f. 2.
flammeum, Pf. t. 142, f. 3.
halophilum, Bn. t. 6, f. 3.
orophilum, Bn. t. 142, f. 4.
semiclausum, Pf. t. 6, f. 2.

CYCLOTUS.[16]

semistriatus,[17] Sow. t. 4, f. 9.
subdiscoideus,[18] Sow. t. 4, f. 10.

DITROPIS.

Bedomei, Bl. t. 136, f. 8, 9
convexus, Bl. t. 136, f. 7, 10.
planorbis, Bl. t. 136, f. 5, 6.

CYCLOPHORUS.[19]

affinis, Th. t. 2, f. 7; t. 48, f. 2; t. 104, f. 1.

[1] J. As. Soc. Beng. 1851, p. 348, pl. 1, f. 3.—Pf. Mon. Pn. vol. 3, p. 53. Kolamullies, S. India.
[2] Id. 1875, vol. 43, pt. 2, p. 147, pl. 3, f. 4.
[3] Id. 1871, vol. 40, pt. 2, p. 93, pl. 5, f. 3. Nongjinghi Hill, Jaintia.
[4] Id. 1874, vol. 43, pt. 2, p. 146, pl. 3, f. 1.
[5] Id. 1874, p. 149, pl. 3, f. 7.
[6] Id. 1874, p. 148, pl. 3, f. 6.
[7] Id. 1874, p. 117, pl. 3, f. 3.
[8] Id. 1874, p. 146, pl. 3, f. 2.
[9] Id. 1875, vol. 44, pt. 2, p. 8, pl. 4, f. 2.
[10] Including, perhaps wrongly, Spiraculum and Rhiostoma. Th. P. Troschel, named by Benson from a mere drawing (An. Nat. H. 1851, ser. 2. vol. 8, pl. 5), which reminds one of the P. pictus of Troschel, was unknown even to Benson himself.
[11] The Alluvei of Benson (not Pf.) proved, when washed, to be a mere form of this very variable species.
[12] A most variable shell, to which Pfeiffer refers the cornu-venatorium of Adams, and to which we must probably assign the f. 12 (not 11) of the Cyclostoma c. v. of Sowerby's Thesaurus.
[13] Malak. Blät. 1867 (vol. 4), p. 85.—Pfeif. Mon. Pneum. vol. 2, p. 39.
[14] The L. Birmanum of Pfeiffer (in Kuster's Chemnitz Cyclos. pl. 47. f. 3, 4, 5) is merely a young Cyclophorus, and the type of the L. subconicum of Reeve has the locality attached of "Cochin China," not Ceylon, as printed.
[15] Benson's type of C. vitreum from the Andamans was a bad specimen of this; compare, likewise, the young with his Helix oeryx (An. Nat. H. 1859, p. 184).
[16] The Malabar locality of the C. spurcus of Grateloup has not been confirmed.
[17] The specimen figured links the less turbinate typical form with the C. montanus of Reeve (C. Icon. Cyclot. f. 58 (as of Pf. Mon. Pn. vol. 2, p. 23).
[18] We include the C. Troilli of Pfeiffer (Proc. Zool. 1862, pl. 12, f. 4).
[19] We have purposely omitted C. cucullatus of Gould (Proc. Bost. N. H. 1856, and Otia, p. 236), the typical specimen figured in Reeve's "Iconica," being a mere monstrosity.

alakastrum, Pf. t. 144, f. 5.
altivagus, Bn. t. 34, f. 2, 3, 6.
annulatus, Tros. t. 143, f. 1, 4.
arthriticus, Th. t. 1, f. 4.
aurantiacus, Schum. t. 33, f. 4.
Aurora, Bn. t. 3, f. 4.
Bairdi, Pf. t. 4, f. 1.
balteatus, Bn. t. 3, f. 1.
Bensoni, Pf. t. 34, f. 5.
cadiscus, Bn. t. 105, f. 10; t. 3, f. 8, as Thwaitesii.
calyx, Bn. t. 4, f. 4.
Ceylanicus,[1] Pf. t. 33, f. 2.
cœlocosus,[2] Bn. t. 4, f. 5.
cornu-venatorium, Bn. t. 104, f. 5, 6.
cratera, Bn. t. 47, f. 8.
cryptomphalos, Bn. t. 3, f. 7.
cytopoma,[3] Bn. t. 47, f. 9.
deplanatus, Pf. t. 3, f. 10.
expansus, Pf. t. 2, f. 3, 4.
exul, Bn. t. 47, f. 10.
eximius, Moss. t. 33, f. 1.
flavilabris, Bn. t. 1, f. 1.
folinseus,[4] Chem. t. 2, f. 5, 6.
fulguratus, Phil. t. 144, f. 1; t. 3, f. 3; t. 3, f. 2?

Haughtoni,[5] Th. t. 1, f. 3; t. 3, f. 6; t. 48, f. 6.
Himalayanus, Pf. t. 34, f. 4.
hispidulus, Bl. t. 47, f. 5, 6.
Indicus, Desh. t. 48, f. 3.
Inglisianus, Stol. t. 143, f. 8, 9.
involvulus, Müll. t. 2, f. 1.
Jerdoni, Bn. t. 33, f. 5, 6.
Layardi, Ad. t. 104, f. 2, 3.
loxostoma,[6] Pf.
Malayanus, Bn. t. 48, f. 4.
Menkeanus, Phil. t. 33, f. 3.
Nilagiricus,[7] Bn. t. 1, f. 5.
opinis, Hanl. t. 144, f. 6.
parapsis,[8] Bn. (Ceylon).
parma, Bn. t. 115, f. 2, 3.
patens,[9] Bl. t. 3, f. 5.
Pearsoni, Bn. t. 48, f. 5 (not t. 1, f. 5[10]).
perniodilis, Gould, t. 1, f. 7.
Phayrei, Th. t. 144, f. 3, 4.
phœnotopicus, Bn. t. 4, f. 3.
pinnulifer, Bn. t. 144, f. 1.
polynema,[11] Pf. t. 2, f. 8.
porphyriticus, Bn. t. 105, f. 4.
pyrotrema,[13] Bn. t. 2, f. 9, 10.

ravalus, Bn. t. 105, f. 5, 6.
scurra, Bn. t. 105, f. 2, 3.
serrutizona, Thorpe, t. 144, f. 7.
Shiplayi, Pf. t. 143, f. 7, 10.
Stanousis, Sow. t. 48, f. 7.
speciosus, Ph. t. 104, f. 4, 7.
stetomphalos, Pf. t. 34, f. 1.
strss-osous, Sow. t. 105, f. 7, 8.
var.? f. 9.
sublaevigatus, Bl. t. 34, f. 7.
subplicatulus, Bedd. t. 115, f. 5, 7.
Theobaldianus, Bn. t. 1, f. 2; var. t. 144, f. 2.
Thwaitesii, Pf. t. 3, f. 9 (not?); tristis, Bl. t. 115, f. 5, 6.
tryblium, Bn. t. 48, f. 1; var. t. 47, f. 10.

OPISTHOMA.
Hindnorum, Bl. t. 6, f. 5, 6.

HELICINA.
Andamanica,[12] Bn. t. 46, f. 10.
Araicanensis, Bl. t. 46, f. 9.
scrupulosa, Bn. t. 133, f. 8, 9.

FRESHWATER SHELLS.

NERITINA.[14]
coluber, Thorp. t. 157, f. 10.

faliginosa, Th. t. 157, f. 8, 9.
obtusa, Bn. t. 157, f. 7, 10.

Perotetiana, Récl. t. 157, f. 2, 3.
reticularis, Sow. t. 157, f. 5, 6.

[1] The reference to Sowerby by Pfeiffer, who at first (in Küster's Chemnitz) attributed the species to Sowerby, was f. 320, 321; the latter was omitted by our printer; the Küster is, however, the less like our ideal.

[2] The fine white basal groove which indents the inner lip is a useful distinguishing characteristic.

[3] Having washed the type, the band alluded to by Benson proved to be mere dirt, and in place of pale flames near the open there were red wavy lines on a paler ground.

[4] The C. Laxi of Tryon (Amer. J. C. vol. 5, p. 111, pl. 10, f. 6), who doubts its identity with the Chemnitzian species.

[5] Reeve, who did not know this shell, quotes it as a synonym of excellens (said to come from Birmah), the single type of which is much nearer his tabicens.

[6] Proc. Zool. Soc. 1852, p. 146; Küst. Chemn. Cycl. pl. 49, f. 12, 13, as Cyclost.; Mon. Pneum. vol. 1, p. 93. Cyclos. Reeve's figure (C. Icon. Cyclop. f. 83) looks very different from Küster's, yet both were from Cuming's types.

[7] Found also in the jungle of Coorg.

[8] Am. Nat. 1855, ser. 2, vol. 12, p. 96.—Pf. Mon. Pneum. vol. 2, p. 66. Reeve's figure is not our ideal.

[9] This species is figured in Pfeiffer's "Novitates," pl. 98, f. 3, 4.

[10] This appears either a form of The dubbanus (?) or possibly an unrecognised species.

[11] Cyclostoma, f. 115, of Sowerby's Thesaurus (vol. 1) seems identical.

[12] In Benson's manuscript I note that he regards this species as identical with the conyntus of Hutton (J. As. Beng. vol. 3, p. 82) from Rajnahal.

[13] According to Sowerby, this is the Megalomeris of Pfeiffer (Pr. Z. 1857, p. 111, and Mon. Pneum. vol. 2, p. 216) so far that same would have priority, but he is wrong.

[14] To limit our work, we have excluded (no estuary species (of course the marine), and, perhaps, a few fluviatile ones, at least such as are found far off from the sea. The omitted shells are N. Peguensis, Bl. (? Baugdensis, Petit?); N. rostrata, Reeve; N. crepidularis, Lam.; N. violacea, Gmel. (depressa, Bn.); N. cornuopia, Bn. (melanostoma, Tros. and Indica Sow.) with which estuary shells may be compared N. pilosba, Bn.; N. intermedia, Desh., and N. nitrata, Reck (in Gal. Douai.; N. Layardi, Reeve; N. Cabotaria, Reck.; N. trisrialis, Sow.; N. Smithii, Gray (tigrina, Benson, hamoligera, Tros.); and N. Coromandeliana of the Conchological Illustrations.

NAVICELLA.
cærulescens, Reeve, t. 137, f. 2, 3.
compressa, Pease. t. 137, f. 1, 4.
Lavesayi, Dohrn, t. 137, f. 8, 9.
reticulata, Reeve, t. 137, f. 5, 6.
squamata, Dohrn, t. 157, f. 1, 4.

ANCYLUS.[1]
Ceylanica, Bn. t. 81, f. 1, 4.
verruca, Bn. t. 81, f. 2, 3.

CAMPTONYX.
Theobaldi, Bn. t. 81, f. 5, 6.

SUCCINEA.[2]
acuminata, Bl. t. 68, f. 7.
Baconi, Pf. t. 68, f. 1, 4.
Bensoni, Pf. t. 67, f. 9.
Ceylanica, Pf. t. 158, f. 10.
collina, Bl. t. 68, f. 8, 9, 10.
crassiuscula, Bn. t. 68, f. 5, 6.
dancina, Pf. t. 67, f. 7.
birmanica, Th. t. 67, f. 5, 6.
Indica, Pf. t. 67, f. 1, 4.
Pfeifferi, Reeus. (Cashmire).
plicata, Bl. t. 67, f. 8.
putris, Lin. (Cashmire).
rutilans, Bl. t. 67, f. 10.
subgranosa, Pf. t. 158, f. 8, 9.
semiserica, Gould, t. 67, f. 2, 3.
vitrea, Pf. t. 68, f. 2, 3.

LITHOTIS.
rupicola, Bl. t. 81, f. 7.
tumida, Bl. t. 81, f. 8, 9.

PHYSA.
(?) Coromandelica,[2] Dunker.

CAMPTOCERAS.
Austeni, Bl. t. 158, f. 3, 4.
lineata, Bl. t. 158, f. 5, 6.
terebra, Bl. t. 158, f. 1, 2.

LIMNÆA.[4]
acuminata, Lam. t. 69, f. 8, 9.
amygdalus, Troseh. t. 69, f. 7, 10.
auricularia,[2] Linnæus.
brevicauda, Sow. t. 158, f. 7.
chlamys,[5] Bn. t. 69, f. 5, 6.
luteola, Lam. t. 70, f. 5, 6.
marginata,[7] Michaud, in Drap. Sup.
ovalis, Gray, t. 70, f. 2, 3, 4.
peregra,[5] Linnæus.
pinguis, Dohrn, t. 70, f. 7, 8, 10.
rufuscens, Gray, t. 69, f. 1, 4; t. 70,
f. 1, 9.
stagnalis,[5] Lin.
tigrina.[6] Dohrn.
truncatula,[5] Müller.

MELANIA.[2]
acanthica, Dohrn (as of Lea), t.
110, f. 19.
laccata, Gould, t. 75, f. 1 to 4.
latana, Gould, t. 74, f. 8, 9.
Broti, Dohrn, t. 71, f. 2, 3.
confusa, Dohrn, t. 72, f. 4.
datura, Dohrn, t. 73, f. 10.

episcopalis, Lea, t. 72, f. 7; t. 75,
f. 5, 7.
fossata, Born, t. 109, f. 4.
gloriosa, Anth. t. 72, f. 1, 2.
Hankeyi, G. A. t. 110, f. 5.
Herculea, Gould, t. 72, f. 5; t. 109,
f. 7.
Hugeli, Phil. t. 71, f. 5, 6.
humerosa,[10] Gould (Tavoy).
Iravadica, Bl. t. 71, f. 1.
jugicostis, Bn. t. 110, f. 8, 9.
Layardi, Dohrn, t. 73, f. 9.
lineata, Gray, t. 71, f. 7.
Menkiana, Lea, t. 110, f. 6.
pagodula, Gould, t. 153, f. 3.
Peguensis,[11] Ant. t. 72, f. 6.
peasmorsa, Tryon, t. 155, f. 2.
pyramis, Bn. t. 110, f. 3, 4.
Reevei, Brot. t. 72, f. 3; t.153, f. 1, 5.
Riqueti, Grat. t. 71, f. 10.
rudis, Lea, t. 74, f. 7, 10.
scalon, Müll. t. 73, f. 1 to 4; var.
elegans, t. 73, f. 5, 6, 7; var.
spinulosa, t. 110, f. 7.
spinata, G. A. t. 109, f. 1.
terebra, Bn. t. 71, f. 8, 9.
tigrina, Hutt. t. 110, f. 1, 2.
Tironri ? Fér. t. 74, f. 5, 6.
tuberculata, Müll. t. 74, f. 1 to 4;
t. 73, f. 8, as Layardi.
variabilis,[12] Bn. t. 109, f. 2, 3, 5, 6;
var. spinosa, t. 75, f. 5, 6.
zonata, Bn. t. 71, f. 4.

---

[1] The A. Baconi of Bourguignat, if distinct, has not been found by our Indian collectors.

[2] Compare with our species the S. rugosa of Pfeiffer (Mon. Hel. vol. 2, p. 527), said to have only two and a half whorls, an extremely short spire, and to come from near Pondicherry.

[3] Mal. Blät. 1862. Two immature specimens of a short-spired Physa were obtained at Quiloa by Benson, but have not the contorted spire of this (Australian?) species.

[4] Kuster's monograph has been vainly studied. Besides sundry from that latitudinarian locality "E. Indies," there is a L. ciliva (? ampullacea) from Bengal, and a L. striata (as of Benson?) from Barrakpore, Mr. W. Blanford has sent us a specimen from our Li, Thibet, of what proves to be the L. acularis of the Conch. Iconica, which is perhaps the Lithemis of Kuster, and possibly the lagenla of Kobelt (Mal. Blät. 1872). It is possibly a distorted peregra. In Wiegmann's Archives for 1837 is a paper by Troschel on the Gangetic species, but the descriptions are much too brief. His peuruus seems our pinguis, var. (f. 8), and most evidently is the peregra of Deshayes (Voy. l'Astrag. Zool. pl. 2, f. 13, 14) from Malabar, which is not the shell so termed in Reeve's "Iconica"; his petala we refer to rufescens; his imperua is unknown to us; his ccraseum, calculata, and noctua, are evidently either ovalis or luteola.

[5] From Afghanistan and Cashmire; too well known to figure. L. Baviniana of Hutton (inadequately described in J. Asi. Beng. 1819, p. 656) was, perhaps, peregra.

[4] L. hians of Sowerby's monograph (C. Icon. Lim. f. 57, a) is the young.

[5] From the Sham Provinces.

[6] The type is just like a streaked form of Benson's L. bulla, we doubt it as a species.

[7] See also M. fluctuosa of Gould (Pr. Bost. N. H. vol. 2, p. 219; Otia, p. 200) from Tavoy, which is too curtly described to be identified. The Trrtanomae Clea: Amerilyi (Paludolina Nassochæa Hanl.) of Benson (An. Nat. H. 1860, p. 258), lives in salkish water.

[10] Proc. Bost. N. H. vol. 2 (1847); Otia, p. 200. "Allied to M. Virginica, Say."

[11] The reference to the American Journal must be expunged: it is only a manuscript species.

[12] Too much latitude has been given to this species. We have figured the variabilis, var. vittata of Theobald (J. Asi. Beng.

1865, vol. 31, pt. 2, pl. 3, f. 1 is in pl. 153, f. 7, and varial lib,
var. turrita of the same (do. f. 6) in pl. 155; f. 6, f. 5 of the
same plate is closely allied to the latter, and perhaps all three
are abnormal forms of M. lecata.

[1] Compare Gould's Paludina petrosa (Pr. Bost. N. H. vol. 1,
p. 144; Otia, p. 191) from Birmah.
[2] Pr. Zool. 1852, p. 127. The worn type is very like our
figure of decussata, but is not that species.
[3] Pr. Zool. 1851, p. 92.
[4] In addition to those indicated A. Dolioides, from S. America,
has been wrongly referred to Bombay by Reeve, and A. Luzonica
of Reeve with A. Samarensis of Philippi, attributed to Ceylon
by Dohrn; these localities have not been confirmed. The A.
zygnoea of Réclus, from Bombay, if not an immature shell is

probably the Amphibola referred to by Blanford as very close to
his A. Fortuna.
[a] Compare the A. hepataria of Reeve (C. Icon. Amp. f. 77).
[a] Conch. Icon. Ampullaria.
[2] Martens assigns to this long-forgotten species the P. fallax
of Frauenfeld and Reeve (C. Icon. Pal. f. 51), and agrees with
Mörch and Troschel as to the identity, also, of Swainson's P.
carinata (Zool. Ill. ser. 1, pl. 98) from the Ganges; this last is
referred by Frauenfeld to Remossii. The type of Reeve's only
moesta (very badly figured) on C. Icon. Pal. f. 22, is, at most a
variety of dissimilis.
[5] Mal. Blat. 1862, p. 40. Possleberry.
[5] Proc. Bost. N. H. vol. 2, Otia, p. 192. Birmah.
[10] See, too, the B. truncata of Souleyet (Voy. Bonite, Zool.

xviii   CONCHOLOGIA INDICA.

goniostoma, Hutt. t. 37. f. 7 (as   VALVATA.   exustus, Desh. t. 39, f. 10; t. 40,
   lutea¹).   piscinalis, Müller. (Cashmire).   f. 10.
inconspicua, Dohrn, t. 37, f. 5, 6.   hyptiocyclos, Bn. t. 99, f. 5, 6, 7.
lravalica, Bl. t. 37, f. 10.   PLANORBIS.   labiatus,⁴ Bn.
Nassa³ Th. t. 37, f. 8, 9.   calathus, Bn. t. 39, f. 1, 2, 3.   Morgniensis, Phil. t. 151, f. 5, 6.
oreala,³ Bn. t. 38, f. 8, 9.   Cantori. Bn. t. 40, f. 1, 2, 3.   rotula, Bn. t. 99, f. 2, 3.
pulchella, Bn. t. 38, f. 5, 6.   carnostus, Bn. t. 39, f. 7, 8, 9.   Sindicus, Bn. t. 40, f. 4, 5, 6.
stenothyroides, Dohrn, t. 38, f. 7. compressus, Hut. t. 99, f. 1, 4. Stehmeri, Dohrn, t. 151, f. 6, 7.
   10.   convexiusculus, Hut. t. 99, f. 8, 9. Trochoideus, Bn. t. 39, f. 4, 5, 6.
travancorica, Bn. t. 38, f. 2, 3.   10.   umbilicalis, Bn. t. 40, f. 7, 8, 9.
                                  elegantulus, Dohrn, t. 151, f. 1, 2, 3. ? zebrinus, Dunk.

BRACKISH - WATER SHELLS.

NEMATURA.²   ² foveolata, Bn. t. 37, f. 3.   ¹ monilifera,¹ Bn. t. 37, f. 4.
Deltæ, Bn. t. 37, f. 2.   ¹ minima,⁶ Sow. t. 37, f. 1.

518, pl. 31, f. 22, 23, 24) from the Ganges, said to have a black   ⁴ Ann. Nat. H. 1850, p. 50. The type, said to have a whitish
peristome. It reminds one of cremnoconus.   rib near its mouth, a character perceptible in several species
   ³ Three species were found on Gray's type-tablet, of which   which do not exhibit the other features required, could not be
the one which agreed the best with his very bold and indefinite   described in Benson's collection. Moradabad.
description was nearer inconspicua.   ⁵ Strictly speaking, this genus should have been excluded
   ¹ Add to the synonymy "J. Asi. Soc. Beng. 1870, vol 39, pt   from our work; our design was at first too comprehensive.
2, p. 192, pl. 18, f. 8."   ⁶ Add "Chilka lake."
   ² The inadequately defined pusilla of Gray (Au. Phil. 1855)   ⁷ Add "Stenothyra nu. Bl. Cont. Mal. pt. 8, pl. 2, f. 16. from
may be comporal.   Port Dalhousie, Bassein, Pegu."

LAND AND FRESHWATER SHELLS

OF

BRITISH INDIA.

PLATE I.

CYCLOPHORUS.

1. **C. flavilabris**, Benson, An. Nat. Hist. ser. 3, vol. 6 (1860), p. 193.—Pfeif. Mon. Pneum. vol. 3, p. 68. Pegu.

2. **C. Theobaldinus**, Benson, An. Nat. Hist. ser. 2, vol. 19 (1857).—Pfeif. Mon. Pneum. vol. 2, p. 47.—Reeve, Conch. Icon. Cyclop. f. 41. Tenasserim Valley; Moulmein; Thayet Myo, Birmah.

3. **C. Haughtoni**, Theobald, Journ. Asi. Soc. Beng. vol. 27 (1859), p. 311. The Farm-caves. Moulmein, Tenasserim.

4. **C. arthriticus**, Theobald, Journ. Asi. Soc. Beng. vol. 33 (1864), p. 246. Hills near the river Pegu.

5 **C. Nilagiricus**, Benson, Malak. Blatt. 1854, p. 83.—Pfeif. Mon. Pneum. vol. 2, p. 52.—Reeve, Conch. Icon. Cycloph. f. 6.—C. Firricanus, Pfeif. Proc. Zool. 1853. Nilgherries; Khoondah Hills.

6. **C. Pearsoni**, Benson, An. Nat. Hist. ser. 2, vol. 8 (1857), p. 185.—Pfeif. M on. Pneum. vol. 2, p. 42 : Kust. ed. Chemn. Cyclos. f. 649.—Reeve, Conch. Icon. Cyclopdt. f. 10. Incot and Chaila, Khasia Hills; Assam.

7. **C. pernobilis**, has been considered identical with the Aurantiacus of Schumacher, but the figures of Chemnitz (1064, 5) suit better the shell delineated by Reeve (Cycl. f. 3), Gould, Boston Jour. Nat. His. 1844, p. 458, pl. 24, f. 11. Mergui; Tavoy.

PLATE II.

CYCLOPHORUS.

1. **C. involvulus**, Muller, Hist. Verm. pt. 2, p. 84 (as Helix).—Pfeif. Mon. Pneumon. vol. 1, p. 59.—Reeve, C. Icon. Cycloph. f. 1. — Turbo volvulus, Chemn. Conch. vol. 9, p. 1066. Ceylon.

2. **C. zebrinus**, Benson, Journ. Asi. Soc. Beng. vol. 5 (1836), p. 355 as (Cyclostoma). — Pfeif. Mon. Pneum. vol. 1, p. 71 ; Kust. ed. Chem. Cyclost. pl. 31, f. 21, 22, 23.—Reeve, C. Icon. Cycloph. pl. 11, f. 46. Nanclai ; Tenasserim.

Was at first erroneously supposed to be the perdix of Sowerby.

3, 4. **C. expansus**, Pfeiffer, Proc. Zool. 1851, p. 242: Mon. Pneum. vol. 1, p. 65 ; Kust. ed. Chem. Cyclos. pl. 39, f. 20, 21.—Reeve, C. Icon. Cycloph. f. 18. Tenasserim.

5, 6. **C. foliaceus**, Chemnitz, Conch. Cab. vol. 9, f. 1069, 1070 (as Turbo).—Pfeif. Mon. Pneumon. vol. 3, p. 65. Andamans.

7. **C. affinis**, Theobald, Journ. Asi. Soc. Beng. vol. 27 (1859), p. 314. Moulmein.

8. **C. polynema**, Pfeiffer, Proc. Zool. 1854, p. 126 ; Mon. Pneum. vol. 2, p. 46. Midnapore, &c.

9. **C. pyrotrema**, var. Benson.

10. **C. pyrotrema**, Benson, An. Nat. Hist. ser. 2,
vol. 14, p. 112.—Pfeif. Mon. Pneum. vol. 2, p. 45.—
Reeve, C. Icon. Cycloph. f. 13.

Sikrigalli, Pathargata, Behar, Rajmahal.

According to Benson was erroneously supposed by
him (Zool. Journ. vol. 5, p. 62) to be the true *incoludes.*

PLATE III.

CYCLOPHORUS.

1. **C. balteatus**, Benson, An. Nat. Hist. ser. 2,
vol. 19 (1857, March): separate pamphlet, p. 7.—
Pfeif. Mon. Pneum. vol. 2, p. 45.

Pegu.

2. **C. arthritic**, var. fulgurans, Theobald.

3. **C. fulguratus**, Pfeiffer, Proc. Zool. 1852, p. 63 ;
Mon. Pneum. vol. 1, p. 80.—Reeve, C. Icon.
Cycloph. f. 35.

Between Thayet Myo and Rangoon.

4. **C. Aurora**, Benson, An. Nat. H. ser. 2, vol. 8 (1851),
p. 186 (as Cyclostoma A.)—Pfeif. Mon. Pneum.
vol. 2, p. 51.—Reeve, C. Icon. Cycloph. f. 36.—
Cyclostoma stenomphalum, var. Pf. in Kust. ed.
Chemn. Cycl. pl. 50, f. 11-13.

Darjiling, Sikkim-Himalaya.

Pfeiffer's original description of *Himalayanus* suits
this shell better than it does the specimen figured by
Mr. Reeve, as that species.

5. **C. patens**, Blanford, Journ. Asi. Soc. Beng.
1862, p. 143.—Pfeif. Mon. Pneum. vol. 3, p. 62.

Thayet Myo, Prome, Henzada, Pegu.

6. **C. Haughtoni**, var. Theobald.

7. **C. cryptomphalus**, Benson, An. Nat. Hist. ser. 2,
vol. 19 (1857), March.—Pfeif. Mon. Pneum. vol. 2,
p. 58.—Reeve, C. Icon. Cyclop. f. 37.

Ava.

8, 9. **C. Thwaitesii**, Pfeiffer, Proc. Zool. Soc. 1854,
p. 127 : Mon. Pneum. vol. 2, p. 66.—Cyclop.
annulatus, var. Thwaitesii, Reeve, C. Icon. Cyclop.
pl. 18, f. 87.

Ceylon.

10. **C. dophianatus**, Pfeiffer, Proc. Zool. 1854, p. 301:
Mon. Pneum. vol. 2, p. 62.—C. annulatus, var.
Reeve, C. Icon.

Khoondah Mountains.

PLATE IV.

CYCLOPHORUS AND ALLIED GENERA.

1. **C. Bairdi**, Pfeiffer, Proc. Zool. 1852, p. 141, pl. 13,
f. 1 : Mon. Pneum. vol. 1, p. 91.

Ceylon.

2. **C. pinnulifer**, Benson, An. Nat. Hist. ser. 2,
vol. 19 (1857).—Pfeif. Mon. Pneum. vol. 2, p. 41 ;
Novit. vol. 1, pl. 37, f. 22, 23, 24.—Reeve, Conch.
Icon. Cycloph. f. 103.

Teria Ghat.

3. **C. phænotopicus**, Benson, An. Nat. Hist. ser. 2,
vol. 8, p. 190, and ser. 2, vol. 10, p. 271.—Pfeif. Mon.
Pneum. vol. 2, p. 54 ; Kust. ed. Chemn. Cyclost.
pl. 50, f. 20, 21.—Reeve, C. Icon. Cyclop. f. 91.

Darjiling, Sikkim Himalayah.

4. **C. calyx**, Benson, An. Nat. Hist. ser. 2, vol. 17,
p. 228.—Pfeif. Mon. Pneum. vol. 2, p. 56 ; Novit.
vol. 1, pl. 37, f. 25, 26, 27.—Reeve, C. Icon.
Cycloph. f. 104.

Moulmein : Akoutong, near the Irrawady,
Birmah.

5. **C. cœloconus**, Benson, An. Nat. Hist. ser. 2,
vol. 8, p. 189.—Pfeif. Mon. Pneum. vol. 2, p. 59 :
Chemn. ed. Kust. Cyclost. pl. 50, f. 9, 10.—Reeve,
C. Icon. Cyclop. f. 89.

Nilgherries.

6. **Aulopoma Itieri**, var.

7. **Aulopoma Itieri**, Guerin, Mag. Zool. 1847, p. 1
(as Cyclostoma).—Pfeif. Mon. Pneum. vol. 1, p. 52.

Ceylon.

8. **Aulopoma Helicinum**, Pfeiffer, Mon. Pneum.
vol. 1, p. 53 (as Turbo II. Chemn. vol. 9, f. 1067, 8).

Ceylon.

9. **Cyclotus semistriatus**, Sowerby (as Cyclostoma),
Proc. Zool. Soc. 1843, p. 29 : Thes. Conch. vol. 1,
p. 91, pl. 23, f. 6.—Pfeif. Mon. Pneum. vol. 1,
p. 22.—Reeve, C. Icon. Cyclot. pl. 4, f. 16.

Poonah, &c.

This is the Cyclostoma fasciatum of Hutton (Journ.
Asi. Soc. Bengal. vol. 3, p. 82).

10. **Cyclotus sub discoideus**, Sowerby, Thes. Conch.
vol. 2, p. 161*, pl. 31, B. f. 304, 305 (as Cyclostoma).—Reeve, C. Icon. Cyclot. pl. 4, f. 21.—Pf.
Mon. Pneum. vol. 1, p. 31.—Cyclostoma aratum,
Benson, An. Nat. Hist. 1851, vol. 8.—Cyclostoma
rusticum, Pfeif. Proc. Zool. Soc. 1851.

Northern Circars.

CONCHOLOGIA INDICA.

PLATE V.

PTEROCYCLOS, INCLUDING SPIRACU-
LUM AND RHIOSTOMA.

1. P. (Sp.) Mastersi, Blanford, MSS.
Assam.
The description will be found in the Zoological Pro-
ceedings, or the Asiatic Journal.

2. P. bilabiatus, Benson, Zool. Journ. vol. 5,—Cyclostoma b. Sowerby, Thes. Conch. vol. 1, p. 110,
pl. 25, f. 81, 82.—Pfeif. Mon. Pneum. vol. 1,
p. 42.—Reeve, Conch. Icon. Pter. pl. 3, f. 12.
Salem in Madras: Behar.

3. P. parvus, Pearson (as Spiraculum p.) Journ.
Asi. Soc. Beng. vol. 2 (1833), p. 392.—Benson,
Journ. Asi. Soc. Beng. vol. 5 (1836), p. 357.—
Pfeif. Mon. Pneum. vol. 1, p. 48.—Reeve, C. Icon.
Pter. pl. 34, f. 15.
Khasia Hills: Assam.
The painting of the Assam variety here figured is
exactly that of the Cingalese P. Cumingii.

4. P. (Sp.) hispidus, Pearson, Journ. Asi. Soc.
Beng. vol. 2, 1833, p. 592, as Spiraculum.—
Benson, Journ. Asi. Soc. Beng. 1836, p. 355, as
Pt.—Cyclos. spiraculum, Sow. Thes. Con. vol. 1,
p. 110, pl. 31, f. 270-272.—Steganotoma Prin-
cepsii, Von dem Busch, in Philip. Abbild. N. Conch.
vol. 1, p. 106, Cycl. pl. 1, f. 6.
Khasia Hills.

5. P. Cingalensis, Benson, An. Nat. Hist. ser. 2,
vol. 11.—Pfeif. Mon. Pneum. vol. 2, p. 29.
Monnlegalla, Ceylon.

6. 7. P. insignis, Theobald, Jour. Asi. Soc. Beng.
1865, vol. 34, pt. 2: separate pamphlet p. 6.
Shan States.

8. P. rupestris, Benson, Journ. Asi. Soc. Beng.
vol. 1, 1832, p. 11, pl. 1, f. 2.—Reeve, C. Icon.
Pteroc. pl. 2, f. 8.—Steganotoma picta, Troschel,
Wiegm. Archiv Naturges, vol. 3, 1837, p. 165,
pl. 3, f. 12, 13, and in Philippi, N. Conch. vol. 1,
Cyclos. pl. 1, f. 5 (teste Pfeif.).
Cuttack.

9. P. Feddeni, Blanford, Journ. Asi. Soc. Beng.
1865, p. 93: Contr. pt. 5.
Thayet Myo, Pegu.

10. P. (Rh.) Haughtoni, Benson, An. Nat. Hist. ser. 3,
vol. 5, p. 96, as Rhiostoma.—Pfeif. Mon. Pneum.
vol. 3, p. 39, as Rh.—P. H. Reeve, Conc. Icon.
Pteroc. f. 30.
Near Moulmein.

PLATE VI.

LEPTOPOMA, OTOPOMA, HELICINA, &c.

1. L. cybeus, Benson, An. Nat. Hist. 1857, ser. 2,
vol. 19. — Pfeif. Mon. Pneum. vol. 2, p. 71:
Novit. pl. 37, f. 28, 29, 30.—Reeve, C. Icon.
Lept. f. 7.
Nautchai.

2. L. semiclausum, Pfeiffer, Proc. Zool. 1851,
p. 202; Mon. Pneum. vol. 2, p. 79.—Reeve, C.
Icon. Lept. f. 35.
Ceylon.

3. L. halophilum, Benson, An. Nat. Hist. ser. 2,
vol. 7, 1851, p. 265 (as Cyclostoma).—Pfeif. Mon.
Pneum. vol. 1, p. 118 (ditto).—Reeve, C. Icon.
Lept. f. 49 (as Lept.).
Point de Galle, and Colombo, Ceylon.

4. L. aspirans, Benson, An. Nat. Hist. ser. 2,
vol. 17, p. 229. — Pfeif. Mon. Pneum. vol. 2.
p. 72.—Reeve, C. Icon. Lept. f. 18.
Tenasserim.

5. Otopoma Hinduorum, Blanford, An. Nat.
Hist. 1864, for O. clausum, Benson.—Pfeif. Mon.
Pneum. vol. 3, p. 122 (as Cyclotus).
Kattiawar, Western India.

6. Otopoma Hinduorum, var.

7. Lagocheilus scissimargo, Benson, An. Nat.
Hist. ser. 2, vol. 17, p. 228 (as Cyclop.?).—Pfeif.
Mon. Pneum. vol. 2, p. 61 (as Cycloph. ?).—Reeve.
C. Icon. Cycloph. f. 105.—Blanf. An. Nat. 1864,
June.
Phie Than, Tenasserim.

8. Lagocheilus tomotrema, Benson, An. Nat. Hist.
1857, ser. 2, vol. 19, (as Cyclop.?).—Pfeif. Mon.
Pneum. vol. 2, p. 50 (as Cycloph.).—Blanford, An.
Nat. Hist. 1864, June.
Teria Ghat, Khasia Hills.

9. Helicina Arakanensis, Blanford, Journ. Asi.
Soc. Beng. 1865, p. 85.
Ramri Island, coast of Arracan.

10. H. Andamanica, Benson, An. Nat. Hist. ser. 3,
vol. 6, 1860, p. 194.—Sow. Thes. Conch. Helic.
pl. 276.
Andaman Islands.

PLATE VII.

MEGALOMASTOMA, PUPINA, POMATIAS.

1. **M. gravidum**, Benson, An. Nat. Hist. ser. 2, vol. 17, 1856, p. 229.—Cyclostoma pollex. Gould. Proc. Boston Soc. N. H. vol. 6, 1856.—Hybocystis g. Pfeif. Mon. Pneum. vol. 3, p. 56.—Pollicaria g. Sow. Thes. Conch. vol. 3, pl. 265, f. 5. Moulmein.

2. **M. funiculatum**, Benson, Journ. Asi. Soc. Beng. vol. 7, 1838, p. 217 (as Cyclostoma).—Pfeif. Mon. Pneum. vol. i. p. 135 (ditto). Darjiling, Himalaya.

3. **M. sectilabrum**, Gould, Boston, J. Nat. His. 1844, p. 459, pl. 21. f. 10 (as Cyclostoma).— Pfeif. Mon. Pneum. vol. 2, p. 81 (ditto). Tavoy ; Yunglaw, Tennasserim.

4. **Pup. arula**, Benson, An. Nat. Hist. ser. 2, vol. 17, p. 230.—Sow. Thes. Conch. vol. 3, pl. 265, f. 3.—Pfeif. Mon. Pneum vol. 2, p. 95. Birmah.

5. **Pup. artata**, Benson, An. Nat. Hist. ser. 2, vol. 17, p. 230.—Pfeif. Mon. Pneum. vol. 2, p. 96.—Sow. Thes. Conch. vol. 3, pl. 265, f. 1. Birmah.

6. **Pup. Blanfordi**, Theobald, Journ. Asi. Soc. Beng. 1864, vol. 33, p. 247. Pegu.

7. **Pup. imbricifera**, Benson, An. Nat. Hist. ser. 2, vol. 17, p. 230.—Pfeif. Mon. Pneum. vol. 2, p. 94.—Sow. Thes. Conch. vol. 3, pl. 265, f. 4, 5. Sylhet, and Teria Ghat.

8. **Pom. Peguensis**, Theobald, Journ. Asi. Soc. Beng. 1864, vol. 33, p. 248. Pegu.

9. **Pom. Himalayæ**, Benson, An. Nat. Hist. ser. 2, vol. 3, 1859, p. 183.—Pfeif. Mon. Pneum. vol. 3, p. 169.—Sow. Thes. Conch. vol. 3, pl. 259, f. 19, 20. Rungun : Darjiling.

10. **Pom. pleurophorus**, Benson, An. Nat. Hist. 1859, ser. 3, vol. 3, p. 183 (previously as Bulimus, 1857).—Bulimus, p. Pfeif. Mon. Helic. vol. 4, p. 474. Teria Ghat ; Khasia Hills.

PLATE VIII.

BOYSIA, HYPSELOSTOMA, AND STREPTAXIS.

1. **B. Bensoni**, Pfeiffer.—Tomogeres Boysii, Pfeif. Symb. pt. 3, p. 82.—Anostoma B. Pfeif. (as of Benson MSS. in Kust. ed. Chemn. Helix, pl. 101, f. 25 to 28. Chittore in Rajpootana, and Ajmere, Bengal.

2. **H. Bensonianum**, Blanford, Journ. Asi. Soc. Beng. vol. 32, p. 326. Mya Leit Doung, S. of Maudalay, Ava.

3. **H. tubiferum**, Benson, An. Nat. Hist. ser. 2, vol. 17, p. 129 (as Tanystoma). Mya Leit Doung, S. of Maudalay, Ava.

4. **S. Petiti**, Gould, Boston J. Nat. Hist. vol. 4, p. 456, pl. 24, f. 7.—Pfeif. Mon. Helic. vol. 1, p. 8.—Philip. Ab. N. Cone. vol. 2, Helix. pl. 8, f. 11. Near Tavoy.

5. **S. Birmanica**, Blanford, J. Asi. Soc. Beng. 1865, p. 81.—Pfeif. Mon. Helic. vol. 5, p. 444. Pegu.

6. **S. Andamanica**, Benson, An. Nat. Hist. ser. 3, vol. 6, 1860, p. 192.—Pfeif. Mon. Helic. vol. 5, p. 444. Andaman Islands.

7. **S. Sankeyi**, Benson, An. Nat. Hist. 1859. Near Moulmein.

8. **S. Watsoni**, Blanford, Journ. Asi. Soc. Beng. vol. 29, p. 126.—Pfeif. Mon. Helic. vol. 5, p. 448. Kossadah Hills ; Nilgherries.

9. **S. Theobaldi**, Benson, An. Nat. Hist. 1859.- Pfeif. Mon. Helic. vol. 5, p. 449. Nandsi, Khasia Hills.

10. **S. Blanfordi**, Theobald, Journ. Asi. Soc. Beng. vol. 33, 1864, p. 245. Aracan Hills, Pegu.

CONCHOLOGIA INDICA.

5

PLATE IX.

SPATHA, PSEUDODON, TRIGONODON, UNIO.

1. **S. Soleniformis**, Benson (as Anodon), Journ. Asi. Soc. Bengal, vol. 5 (1836, Nov.), p. 749. Cachar.

If this shell is a true Mycetopus, its specific name must be changed, as D'Orbigny published his M. Soleniformis in 1835 (Guerin's Mag. Zool.); it closely approaches the M. (?) emarginatus of Lea.

2. **P. inocularis**, Gould, Proc. Bost. Soc. Nat. Hist. vol. 1, p. 160 (as Anodon), 161. — Unio Vondembuschiana, Reeve, Conch. Icon. Unio, l. 518, badly (from this specimen).
River Salwen, Birmah.

May possibly prove identical with the Margaritana Vondembuschiana (Lea, Trans. Amer. Phil. Soc. vol. 5, pl. 18, f. 39,) a supposed Javanese species.

3. **T. crebristriatum**, Anthony, Amer. Journ. Conc. vol. 1 (1865), pl. 18, f. 1 (as Monocondylea).— Unio c. Reeve, Conch. Icon. Unio, f. 517.
British Birmah.

4. **P. Salwenianum**, Gould, Proc. Bost. Soc. Nat. Hist. vol. 1, p. 158, 160 (as Anodon and Pseudodon) ; Otia, p. 193.—Unio S. Reeve, Conch. Icon. Unio, f. 513.
Salwen River, Birmah.

5. **T. crebristriatum**, var. Anthony.
The Monocondylea Papuensis of Anthony (Amer. Conch. Journ. vol. 1) seems a form of this rather variable species.

6. **U. lamellatus**, var. Lea.
A peculiar winged form (like Wahlamatensis of Lea) from Mandelay.

Judging from the examination of many hundred specimens of Indian Unionidæ, confirmed by long and peculiar study of thousands of North American Naiades, we are compelled to believe that the teeth and muscular impressions of each species are so liable to modification through age and casualties as scarcely to present any permanent defining character. Not but that they merit examination, as the vividly green variety of tricuboides can instantly be discriminated from corrugatus by its dentition.

According to our notions cæruleus, marginalis, faci-

lens, corrugatus have been unnecessarily subdivided, and several species founded on diseased or eroded shells; but as our aim is to indicate the names which have been applied to the several forms, we shall proceed to figure the more striking ones, never erasing a species from our lists without having traced it to its rest by indisputable links.

PLATE X.

UNIO.

1. **Unio olivarius**, Lea, Trans. Amer. Phil. Soc. ser. 2, vol. 1 (Obs. on Unio, vol. 1, p. 143), pl. 16, f. 38.—Hanley, Recent Bivalve Sh. p. 195, pl. 22, f. 32.—Kuster, ed. Chem. Unio, pl. 82, f. 2.
Cashleund Streams.

2, 1. **U. nucleentus**, Benson, An. Nat. Hist. ser. 3 (1862), p. 187.—U. occatus, Lea, Journ. Ac. Nat. Sc. Philad. 1865 (Obs. on Unio, vol. 10), pl. 50, f. 301.—Reeve, C. Icon. Unio, f. 412.
" Bengal;" Mandelay.

The somewhat worn condition of the type (figure 2) prevented Lea from identifying the species.

3. **U. radula**, Benson, in Hanley's Recent Bivalve Sh. p. 382, pl. 23, f. 44.
Assam.

Benson's unique type (with part of the ferruginous coating removed), from which Mr. Hanley drew up his description, has been once more figured.

5. **U. smaragdites**, Benson, An. Nat. Hist. ser. 3, vol. 10 (1862), p. 190.
Burlumpooter River, Assam.

So rare a shell that the original types are the only examples known to us.

6. **U. Bonneaudi**, Eydoux and Souleyet, Guerin's Mag. de Zool. 1838, pl. 119, f. 1.
Bhamo, Upper Birmah (Blanford).

A very scarce shell, lent to us by Mr. Blanford, whose incomparable series of Indian Uniones was most liberally placed at our disposal.

7. **U. pugio**, Benson, An. Nat. Hist. ser. 3, vol. 10 (1862), p. 190, Reeve, C. Icon. Unio, f. 516.
Ava; Pegu (Theobald).

A rare shell which cannot be confused with any other of the known Indian Uniones.

PLATE XI.

UNIO.

1. **U. favidens,** Benson, An. Nat. Hist. ser. 3, vol. 10 (1862), p. 188, for Glean. Sci. Calcutta, vol. 1, pl. 7, f. 1 (without name).—Reeve, Conch. Icon. Unio, f. 131.

Sunderbund, Bengal.

The species was not actually described until September, 1862.

2. **U. favidens,** var. pinax.—U. pinax, Benson, An. Nat. Hist. ser. 3, vol. 10 (1862), p. 192.

Gungun Stream, near Moradabad, Rohilcund.

Figured from the worn type lent by Mr. Benson.

3. **U. favidens,** var. plagiosoma.—U. plagiosoma, Benson, An. Nat. Hist. ser. 3, vol. 10 (1862), p. 191.

River Cane, near Banda, Bundelcund.

The beaks are eroded in the Bensonian type here figured.

4. **U. Sikkimensis,** Lea, Proc. Philad. 1859, p. 151; Obs. Unio, vol. 7, p. 69, pl. 39, f. 131.—Reeve, Conch. Icon. Unio, f. 400.

Sikkim.

This is in a better condition than Mr. Hanley's original type of the species.

5. **U. crispisulcatus,** Benson, An. Nat. Hist. ser. 3, vol. 10 (1862), p. 193.—Reeve, Conch. Icon. Unio, f. 262.

Tenasserim; Pegu.

6. **U. trirostris,** Benson (not Reeve), in Hanley's Phot. Conch. pl. 1 (1862).

Moradabad.

This well marked form was distributed by Mr. Benson as his U. trirostris (a name, like very many other of his manuscript species) never actually published by himself. The specimen delineated in the Photographic Conchology had been previously sent to Dr. Lea, who regarded it as new to science. It may possibly prove a mere variety of triembolus or favidens.

PLATE XII.

UNIO.

1. **U. pachysoma,** Benson, An. Nat. ser. 3, vol. 10 (1862, Sept.): separate, p. 3.

Berlimapooter River, Assam.

A much more solid and rare shell than the allied cæruleus. In England the only known specimens are in the collections of Benson and Hanley. The Calcutta specimen referred to by Benson is merely a cæruleus.

2. **U. Gerbidoni,** Eydoux and Souleyet, Guerin's Mag. de Zool. 1838, pl. 118, f. 2.

"Coromandel."

Almost runs into cæruleus; in characteristic examples, however, the anterior side is longer in proportion. It is not common.

3. **U. cæruleus,** Lea, Trans. Amer. Phil. Soc. vol. 4 (Obs. on U. vol. 1, p. 165) pl. 13, f. 25.—Hanley, Recent Bivalve Sh. p. 194, pl. 20, f. 49.—Reeve, C. Icon. Unio, f. 230.

River Hooghly, &c. &c.

4. **U. Gerbidoni,** variety, Eyd. and Soul.

5. **U. theca,** Benson, An. Nat. Hist. ser. 3, vol. 10 (1862, Sept.): separate, p. 3.

River Cane, near Banda, Bundelkund.

Our drawing is taken from the unique example in Mr. Benson's collection. The specimen seems scarcely mature, and bears some small resemblance to the young of one of the marginalis group.

6. **U. leioma,** Benson, An. Nat. Hist. ser. 3, vol. 10 (1862, Sept.), p. 192: separate, p. 9.

Near Bombay.

We have figured two shells, one the aged and eroded type of Benson, the other with the apical sculpture perfect. It approaches cæruleus, &c., very closely.

PLATE XIII.

HELIX: Section **Plectopylis**.

1. H. achatina, Gray MSS. in Pfeif. Zeitsch. Mal.
1845, p. 86 : Mon. Helic. vol. 1, p. 406, and vol. 5,
p. 395.
   Farm-caves, near Moulmein.

2. H. plectostoma, Benson, J. Asi. Soc. Beng. vol. 5
(1836), p. 351.—Pfeif. Mon. Helic. vol. 1, p. 415;
vol. 5, p. 417.
   Darjiling, and Klassia Hills.

3. H. perareta, Blanford, J. Asi. Soc. Beng. 1865,
p. 75 (as Plectop.): Cont. pt. 5, p. 11.—Pfeif. Mon.
Helic. vol. 5, p. 397.
   Near Ava.

4. H. repercussa, Gould, Proc. Boston Soc. N. H.
vol. 6, 1856, p. 11 : Otia Conch. p. 219.—Pfeif.
Mon. Helic. vol. 5, p. 396.
   Tavoy and Mergui, in Birmah.

5. H. pinacis, Benson, An. Nat. Hist. ser. 3, 1859,
April, p. 268.—Pfeif. Mon. Helic. vol. 5, p. 417.
   Sikkim (Rungun, and near Pankabari).

6. H. Karenorum, Blanford, Journ. Asi. Soc. Beng.
1865, p. 73 (as Plectop.): Contr. Ind. Mal. pt. 5,
p. 9.
   Pegu.

7. H. anguina, Gould, Journ. Boston Soc. Nat. His.
vol. 2 (1847), p. 218.
   Manko, near Newville, Tavoy, Birmah.

8. H. leiophis, Benson, An. Nat. His. ser. 3, vol. 5
(1860), p. 246 (as Plectop.)—Pfeif. Mon. Helic.
vol. 5, p. 396.
   Thayet Myo.

9. H. refuga, var. dextrorsa, Philippi, Ab. N. Conc.
vol. 3, Hel. pl. 10, f. 4, as of Gould, Proc. Bost.
Phie Than, Tenasserim.

10. H. cyclaspis, Benson, An. Nat. 1859, ser. 3,
vol. 3, p. 273 (for catinus, Benson, not Pfeiffer),
p. 185.
   Near Moulmein, Tenasserim.

PLATE XIV.

HELIX.

1. H. Charpentieri, Pfeiffer, Proc. Zool. Soc. 1853,
p. 127 : Mon. Hel. vol. 4, p. 296.—Reeve, C. Icon.
Hel. f. 1285.
   Ceylon.

2. H. Rivolii, Deshayes, Encyc. Méthod. Vers. vol. 2,
p. 208 : Fer. Hist. vol. 1, p. 7.—Pfeif. Mon. Helic.
vol. 4, p. 297.—Reeve, C. Icon. Helix, pl. 185,
f. 1281.
   Ceylon.

3. H. erronea, Albers, in Pfeif. Mon. Helic. vol. 5,
p. 239.—Reeve, C. Icon. Hel. f. 413.—Pfeif. Mon.
Helic. vol. 4, p. 298.
   Ceylon.

4. H. gabata, Gould, Proc. Bost. Soc. Nat. Hist. and
Journ. Bost. Soc. Nat. Hist. vol. 4, p. 154, pl. 24,
f. 9.—Pfeif. Mon. Helic. vol. 1, p. 396 : vol. 3,
p. 253.
   Birmah.

5. H. capitium, Benson, An. Nat. Hist. 1848 (ser. 2,
vol. 2), p. 160.—Pfeif. Mon. Hel. vol. 3, p. 220.—
Reeve, C. Icon. Hel. f. 749.
   Sicrigully, Behar, N. India.

6. H. hariola, Benson, An. Nat. H. ser. 2, vol. 18,
p. 251.—Pfeif. Mon. Helic. vol. 4, p. 260, and
vol. 5, p. 337 : Novit. vol. 1.
   Birmah.

7. H. Merguiensis, Philippi, Zeitsch. Mal. 1846,
p. 192.—Pfeif. Mon. Helic. vol. 1, p. 397.—Reeve.
C. Icon. Helix. f. 1205.
   Mergui, Birmah.

8. H. bifoveata, Benson, An. Nat. ser. 2, vol. 18.
p. 251.—Pfeif. Mon. Hel. vol. 4, p. 296.
   Tenasserim.

9. H. delibrata (?), var. Benson.—Runs into Zoroaster.

10. H. delibrata, Benson, Journ. Asi. Soc. Beng.
1836, vol. 5, 1836, p. 352.—Pfeif. Mon. Helic.
vol. 1, p. 369.—H. procumbens, Gould, Boston
Journ. Nat. Hist. vol. 4, p. 455, pl. 24, f. 1.—
Reeve, C. Icon. Hel. f. 435.
   Moulmein and Tavoy, Birmah.

Gould's procumbens is said to be identical with the
earlier published delibrata, a species whose brief description was certainly not suggestive of this shell.

PLATE XV.

HELIX.

1. **H. Phayrei**, Theobald, J. Asi. Soc. Beng. 1859, vol. 28, p. 310 : Des. Burm. p. 2.
   Near the Irawady, between Prome and Ava.

2. **H. pylaica**, Benson, An. Nat. H. ser. 2, vol. 18, p. 249.—Pfeif. Mon. Helic. vol. 4, p. 164.
   Birmah.

3. **H. Tickelltii**, Theobald, J. Asi. Soc. Beng. 1859, vol. 28, p. 310 : Desc. Burm. p. 2.—Pfeif. Monog. Helic. vol. 5, p. 267.
   Near Moulmein.

4. **H. Akoutongensis**, Theobald, J. Asi. Soc. Beng. vol. 28, 1859, p. 310 : Desc. Burm. p. 2.—Pfeif. Mon. Helic. vol. 5, p. 408.
   Akoutong, Pegu, near the banks of the Irawady.

5. **H. rotatoria**, Theobald (as of Von dem Busch), J. Asi. Soc. Beng. 1859, vol. 27, p. 317 : Notes on distrib. Shells India, pt. 1. p. 5.
   Akoutong. Pegu.
   As some have doubted the identity of this with the Javanese species, it is merely delineated as the shell so termed by Indian collectors.

6. **H. tapoina**, Benson, J. Asi. Soc. Beng. vol. 5, 1836, p. 352.—An. Nat. Hist. vol. 9, p. 186.— Pfeif. Mon. Helic. vol. 4, p. 207 ; vol. 3, p. 204.— Reeve, Conch. Icon. Helix. f. 750 (from Benson).
   E. frontier of Bengal.

7. **H. Oldhami**, Benson, An. Nat. Hist. ser. 3, vol. 3 (1859), p. 185.—Pfeif. Mon. Hel. vol. 5, p. 410 : Mal. Blät. 1859, p. 31.
   Birmah.

8. **H. Huttoni**, Pfeiffer, Symb. pt. 3 : Mon. Hel. vol. 1, p. 202 (for H. orbicula, Hutton, J. Asi. Soc. Beng. vol. 7, pt. 1, 1848, p. 217, not D'Orb.).— Reeve, C. Icon. Hel. f. 786.
   Simla and Mahassa, Himalaya.

9. **H. Atkinsoni**, Theobald, J. Asi. Soc. Beng. 1859, vol. 28, p. 309 : Desc. Burm. p. 1.—Pfeif. Mon. Helic. vol. 5, p. 177.
   Near Moulmein.

10. **H. tapoina**, Benson, var. Arakanensis.
    H. Arakanensis, Theobald, MSS. (no description), J. Asi. Soc. Beng.
    Aracan.

PLATE XVI.

HELIX.

1. **H. bascauda**, Benson, An. Nat. Hist. 1859, ser. 3, vol. 3, p. 186.—Pfeif. Mon. Helic. vol. 5, p. 256.
   Teria Ghat, Khasia Hills.

2. **H. gratulator**, Blanford, J. Asi. Soc. Beng. 1865, p. 72 : Cont. Mal. pt. 5, p. 8.—Pfeif. Mon. Helic. vol. 5, p. 94.
   Irawady Valley, Pegu.

3. **H. cassidula**, Benson, An. Nat. Hist. 1859, ser. 3, vol. 3, p. 186.—Pfeif. Mon. Helic. vol. 5, p. 75.
   Moulmein.

4. **H. aspirans**, W. and H. Blanford, J. Asi. Soc. Beng. 1861, p. 355, pl. 1, f. 12.—Pfeif. Mon. Helic. vol. 5, p. 84.
   Near Pykara, Nilgherries.

5. **H. fastigiata**, Hutton, J. Asi. Soc. Beng. vol. 7, pt. 1, p. 217.—Pfeif. Mon. Helic. vol. 3, p. 40 : Chemn. ed. Kuster, Helix. p. 144, f. 15, 16.
   Simla and Landour, Himalaya.

6. **H. confinis**, Blanford, J. Asi. Soc. Beng. 1865, p. 71 (as Nanina): Cont. Mal. pt. 5, p. 7.—Pfeif. Mon. Helic. vol. 5, p. 83.
   Thayet Myo, confines of Birmah.

7. **H. polypleuris**, Blanford, J. Asi. Soc. Beng. 1865, p. 76 : Cont. Mal. pt. 5 (1865), p. 12.—Pfeif. Mon. Helic. vol. 5, p. 136.
   Aracan Hills.

8. **H. calpis**, Benson, An. Nat. Hist. 1859, p. 258.— Pfeif. Mon. Helic. vol. 5, p. 64.
   Rungun Valley, near Darjiling.

9. **H. Poongee**, Theobald, J. Asi. Soc. Beng. 1859, vol. 28, p. 311 : Desc. Burm. p. 2.
   Near Moulmein.

10. **H. tortiana**, Blanford, J. Asi. Soc. Beng. 1861, p. 355, pl. 3, f. 11 : Cont. Mal. pt. 2.—Pfeif. Mon. Helic. vol. 5, p. 71.
    Pykara and Neddiwuttum, Nilgherries.

CONCHOLOGIA INDICA

PLATE XVII.

ACHATINA : Section Electra.

1. **A. nitens**, Gray, An. Philos. 1825, p. 415 : Spic. Zool. pd. 6, f. 18.
Ceylon.

2. **A. inornata**, Pfeiffer, Proc. Zool. Soc. 1851 : Monog. Helic. vol. 3, p. 190.
Ceylon.

3. **A. inornata**, var. Pfeiffer

4. **A. Ceylanica**, Pfeiffer, Monog. Helic. vol. 2, p. 258 (not Reeve).—Philippi, Abbil. N. Con. vol. 2, p. 213, Ach. pl. 1, f. 3. — A. oxophila, Benson, in Reeve, Conch. Icon. Ach. f. 105.
Nilgherries.

5. **A. Theobaldi**, Hanley, MSS.
Near the Salwen.
Differs from A. Cassiaca, of which it has been considered a variety, by its smoothness, more convex whorls, &c.

6. **A. prælustris**, Benson, An. Nat. Hist. ser. 3, vol. 5 (1860), p. 462. — Pfeif. Monog. Helic. vol. 6, p. 222.
Orissa.

7. **A. prælustris**, var. Benson.

8. **A. Chessoni**, Benson, An. Nat. Hist. ser. 3, vol. 5 (1860), p. 462.—Pfeif. Monog. Helic. vol. 6, p. 222.
Mahabaleshwar Hills.

9. **A. Tamulica**, W. and H. Blanford, Journ. Asi. Soc. Beng. 1861, p. 362.—Pfeif. Monog. Helic. vol. 6, p. 232.
Cullagoody, near Trichinopoly.

10. **A. textilis**, Blanford, Journ. As. Soc. Bengal, 1866, p. 41.—Pfeif. Mon. Helic. vol. 6, p. 220.
Anamullay Hills.

PLATE XVIII.

ACHATINA : Section Electra.

1. **A. scrutillus**, Benson, An. Nat. Hist. ser. 3, vol. 5 (1860), p. 463.—Pfeif. Mon. Helic. vol. 6, p. 227.
Orissa; Nerbudda.

2. **A. corrosula**, Pfeiffer, Proc. Zool. Soc. 1856, p. 35; Monog. Helic. vol. 4, p 412; Novit. vol. 1, pl. 29, f. 9, 10.
Nilgherries.

3. **A. Fairbankii**, Benson, An. Nat. Hist. (1865, Jan.) p. 14.—Pfeif. Monog. Helic. vol. 6, p. 232.
Mahabaleshwar Hills.

4. **A. hastula**, Benson, An. Nat. Hist. ser. 3, vol. 5 (1860), p. 461.—Pfeif. Monog. Helic. vol. 6, p. 235.
Darjiling.

5. **A. pertenuis**, Blanford, Journ. As. Soc. Beng. 1865, p. 79.—Pfeif. Monog. Helic. vol. 6, p. 237.
Bassein, Pegu.

6. **A. pyramis**, Benson, An. Nat. Hist. 1860, p. 463.—Pfeif. Monog. Helic. vol. 6, p. 226.
Khasia Hills.

7. **A. Orobia**, Benson, An. Nat. Hist. ser. 3, vol. 5 (1860), p. 460. — Pfeif. Monog. Helic. vol. 6, p. 224.
Darjiling.

8. **A. Orobia**, var. Benson.

9. **A. lyrata**, Blanford, MSS. (from type).

10. **A. brevis**, Pfeiffer, Proc. Zool. Soc. 1861, p. 387 : Monog. Helic. vol. 6, p. 227.
Ahmednuggur.

PLATE XIX.

SPIRAXIS AND BULIMUS.

1. **S. Haughtoni**, Benson, An. Nat. Hist. (1863,
Feb.).—Pfeif. Monog. Helic. vol. 6, p. 189.

   Port Blair, Andamans.

2. **B. candolaris**, Pfeiffer, Proc. Zool. Soc. 1846,
p. 40 : Monog. Helic. vol. 2, p. 427.—Reeve, Conch.
Icon. Bulim. f. 408.

   Takht i. Suliman, Cashmire.

3. **B. Kunawarensis**, Hutton, in Reeve Conch. Icon.
Bul. pl. 62, f. 426.—Pfeif. Monog. Helic. vol. 3,
p. 349.

   Landour, W. Himalaya.

4. **B. prætermissus**, Blanford, Journ. Asi. Soc.
Beng. 1861, p. 361 : Cont. Ind. Mal. pt. 2.

   Orissa.

   A large var. which has the freckles of Mavortius.

5. **B. atricallosus**, Gould, Boston Journ. N. H.
vol. 4, p. 457, pl. 24, f. 3.

   Tavoy.

6, 8. **B. intermedius**, Pfeiffer, Proc. Zool. Soc. 1834,
p. 291 : Monog. Helic. vol. 4, p. 386 : Novit.
vol. 1.

   Ceylon.

7. **B. Sikkimensis**, Benson, in Reeve Conch. Icon.
Bul. f. 651 (fry only).

   Darjiling, Sikkim Himalaya.

9. **B. Sylheticus**, Reeve, Conch. Icon. Bulim. f. 564.
—Pfeif. Monog. Helic. vol. 3, p. 322.

   Sylhet, E. Himalayah.

10. **B. Theobaldianus**, Benson, An. Nat. Hist.
(1857, April) : separate, p. 3.—Pfeif. Monog. Helic.
vol. 4, p. 473.

   Yunghaw, Tenasserim.

PLATE XX.

BULIMUS.

1. **B. orbus**, W. and H. Blanford, Journ. Asi. Soc.
Bengal, 1861, p. 15, pl. 1, f. 14.—Pfeif. Monog.
Helic. vol. 6, p. 150.

   Cuttygoody, near Trichinopoly.

2. **B. arcuatus**, Hutton, in Pfeif. Monog. Helic.
vol. 2, p. 118.—Pfeif. Kust. ed. Chemn. Conch. Bul.
pl. 17, f. 1, 2.—Reeve, Conch. Icon. Bul. f. 478.

   Mahassa ; Moulmein.

3. **B. Smithei**, Benson, Ann. Nat. Hist. 1865, Jan.
p. 15.—Pfeiffer, Monog. Helic. vol. 4, p. 56.

   Banks of Sutlej, Punjaub.

   Identified from the type : the shape was incorrectly
described.   It has the general aspect of Nilagiricus.

4. **B. rufistrigatus**, var. gracilis, Benson, in Reeve
Conch. Icon. Bul. f. 570.

   Between the rivers Jumna and Sutlej, Punjaub.

5. **B. vibex ?**, var., Hutton, in Pfeif. Mon. Helic.
vol. 2, p. 118, and Reeve Conch. Icon. Bul. f. 209.

   W. Himalaya.

6. **B. Sindicus**, Benson, in Reeve Conch. Icon. Bul.
f. 303.—An. Nat. Hist. ser. 3, vol. 5 (1860),
p. 464.

   Darjiling.

7. **B. arcuatus**, var. Hutton.

8. **B. salsicola**, Benson, An. Nat. Hist. 1857, April :
separate, p. 1.— Pfeif. Monog. Helic. vol. 4,
p. 423.

   Punjaub.

9. **B. Fairbanki**, Pfeiffer, Proc. Zool. Soc. 1857,
p. 159 : Monog. Helic. vol. 4, p. 410.

   Ahmednuggur.

10. **B. punctatus**, Anton, in Pfeif. Monog. Helic.
vol. 2, p. 212.—Reeve, Conch. Icon. Bul. f. 452.

   Bundelkund, and Southern India.

PLATE XXI.

BULIMUS.

1. **B. Abyssinicus**, Ruppell, in Pfeif. Mon. Helic. vol. 2, p. 110.—Reeve, Conch. Icon. Bulim. f. 296.

Malwah, Central India.

2. **B. Ceylanicus**, Pfeiffer, Symbol. pt. 3, p. 83: Mon. Helic. vol. 2, p. 59.—Reeve, Conch. Icon. Bulim. pl. 43, f. 274.

Ceylon.

3. **B. trifasciatus**, Bruguiere, Encycl. Méth. Vers, p. 317.—Pfeif. Mon. Helic. vol. 2, p. 58.—Reeve, Conch. Icon. Bulim. f. 237.

Ceylon.

The B. fuscoventris of Benson (An. Nat. Hist. vol. 18, p. 96), is this species in a younger stage. We assert this after careful examination of the type.

4. **B. Moussonianus**, Petit, Journ. Conch. 1851, vol. 3, p. 266, pl. 7, f. 4.

Near Bombay.

5, 6. **B. Sinensis**, Benson, An. Nat. Hist. 1851.— Pfeif. Mon. Helic. vol. 3, p. 320.

Sheoray Ghoom, Pegu.

7. **B. Jerdoni**, Benson, in Reeve's Conch. Icon. Bulim. f. 297.—Pfeif. Mon. Helic. vol. 3, p. 335: Kust. ed. Chemn. Bulim. pl. 20, f. 11, 12.

Deccan.

8. **B. albizonatus**, Reeve, Conch. Icon. Bulim. pl. 81, f. 604.

Ceylon.

9. **B. physalis**, Benson, An. Nat. Hist. 1857.—Pfeif. Mon. Helic. vol. 4, p. 386.

Khoonda Ghat, Nilgherries.

10. **B. rufopictus**, Benson, An. Nat. Hist. 1856, p. 96.—Pfeif. Mon. Helic. vol. 4, p. 404.

Ceylon.

PLATE XXII.

BULIMUS.

1. **B. Chion**, Pfeiffer, Proc. Zool. Soc. 1856, p. 332; Mon. Helic. vol. 4, p. 465.

Punjaub.

2. **B. vicarius**, Blanford, J. Asi. Soc. Beng. 1870, vol. 39, pt. 2, p. 18, pl. 3, f. 15.

Habiang in Garo Hills.

3. **B. stalix**, Benson, An. Nat. Hist. 1863, May.— Pfeif. Mon. Helic. vol. 6, p. 61.

Ceylon.

4. **B. Estellus**, Benson, An. Nat. Hist. 1857.—Pfeif Mon. Helic. vol. 4, p. 462.

Simle.

5. **B. domina**, Benson, An. Nat. Hist. 1857.—Pfeif. Mon. Helic. vol. 4, p. 425.

Cashmire.

6. **B. Boysianus**, Benson, in Reeve's Conch. Icon. Bulim. f. 575.

Kemaon, W. Himalayah.

7. **B. Pertica**, Benson, An. Nat. Hist. 1857.—Pfeif. Mon. Helic. vol. 4, p. 462.

Simle.

8. **B. vibex**, var. Hutton and Benson. See plate 23, figure 2.

9. **B. nivicola**, Benson, in Reeve's Conch. Icon. Bulim. f. 496.

Liti Pass, W. Himalayah.

10. **B. insularis**, Ehrenburg, Symbol. Phys. (as Pupa).—Bul. i. Pfeif. in Kust. ed. Chemn. Bulim. pl. 36, f. 26, 27, 28. B. pullus, Gray, Proc. Zool. Soc. 1834.—Reeve, Conch. Icon. Bulim. f. 476.

Delhi, Bundelkund, &c.

PLATE XXIII.

BULIMUS.

1. **B. Agrensis,** Kurr, Mal, Blät. vol 2, 1855, p. 107.
—Pfeif. Mon. Helic. vol. 4, p. 463.
Agra.

2. **B. vibex,** Hutton MSS. in Pfeif. Mon. Helic.
vol. 2, p. 118; Kust. ed. Chemn. Bulim. pl. 17,
f. 5, 6.—Reeve, Conch. Icon. Bulim. f. 299.
Simla and Landour.

3. **B. Nilagiricus,** Pfeiffer, Proc. Zool. Soc. 1846,
p. 418; Mon. Helic. p. 119.—Reeve, Conch. Icon.
Bulim. f. 294.
North Khasi Hills : Nilgherries.

4. **B. gracilis,** Hutton, J. Asi. Soc. Beng. vol. 3,
p. 84, 93.—Pfeif. Mon. Helic. vol. 2, p. 157.—
Reeve, Conch. Icon. Bulim. f. 495.
Bundelkund, and the plain provinces of Bengal,
&c. Ceylon. Birmah.

5. **B. Griffithi,** Benson, in Reeve's Conch. Icon.
Bulim. f. 502.
Afghanistan.

6. **B. eremita,** Benson, in Reeve's Conch. Icon. Bulim.
pl. 78, f. 573.
Bolan Pass, Afghanistan.

7. **B. pretiosus,** Cantor MSS. in Reeve, Conch.
Icon. Bulim. f. 619.
Mimosa bushes, banks of Jhelum, Chillianwalla.
Closely allied to the last species.

8. **B. spelæus,** Hutton. Journ. Asi. Soc. Beng. vol.
18, pt. 2 (1849), p. 655.
Bolan Pass, Afghanistan.
Closely allied to the two preceding, and considered
by Benson identical with his eremita.

9. **B. cœnopictus,** Hutton J. Asi. Soc. Beng. vol. 3,
p. 85, no. 9, and p. 93, and vol. 18, pt. 2, 1849,
p. 654 (as Pupa).—Reeve, Conch. Icon. Bulim.
f. 492.—Pfeif. Mon. Helic. vol. 3, p. 549.
Afghanistan, &c. : Upper Birmah.

10. **B. rufistrigatus,** Benson, in Reeve's Conch. Icon.
Bulim. f. 570.
Between the rivers Jumna and Sutlej.
A somewhat doubtful species, which runs into pre-
tiosus and eremita.

PLATE XXIV.

CLAUSILIA.

1. **C. bacillum,** Benson MSS. in Theobald's Notes
on Distribution, in J. Asi. Soc. Beng.
Nanclai, Khasi Hills.
No description appeared of this species, of which two
specimens alone are known—the latter, or type, of which
is here delineated : the other is in the Bensonian
collection.

2. **C. insignis,** Gould, Proc. Bost. Soc. N. H., and
Journ. Boston Soc. Nat. Hist. vol. 4, p. 458, pl 24,
f. 8, from which Pfeif. Mon. Helic. vol. 2, p. 423.
Tavoy, Birmah.

3. **C. insignis,** var. gracilior, Pfeiffer (as of Gould),
Mon. Helic. vol. 3, p. 589 : Novit. Conch. vol. 1.
Moulmein, Birmah.

4. **C. cylindrica,** Gray, MSS. in Pfeif. Symbol. and
Mon. Helic. vol. 2, p. 428 ; vol. 3, p. 590 ; Kust.
ed. Chemn. Claus. pl. 11, f. 12 to 16.—C. elegans,
Hutton MSS.
Landour.

5. **C. bulbus,** Benson, An. Nat. Hist. 1863, May,
p. 321.—Pfeif. Mon. Helic. vol. 6, p. 410.
Near Moulmein, Birmah.

6. **C. fusiformis,** Blanford, J. Asi. Soc. Beng. 1865,
p. 80.—Pfeif. Mon. Helic. vol. 6, p. 440.
Aracan Hills.

7. **C. loxostoma,** Benson, J. Asi. Soc. Beng. vol. 5
(1836), p. 353.—Pfeif. Mon. Helic. vol. 2, p. 404.
—C. Bengalensis, Von dem Busch's MSS. in Kust.
ed. Chemn. Claus. pl. 2, f. 11—13, and Pfeif. Mon.
Helic. vol. 2, p. 60.
Teria Ghat.

8. **C. Masoni,** Theobald, J. Asi. Soc. Beng. vol. 33
(1864), p. 246 ; separate pamphlet, p. 16.
Near Tonghoo, in the mountains between Pegu
and Martaban.

9. **C. tuba,** Hanley, Proc. Zool. Soc. 1868.
Shan States (Theobald).
At present of extreme rarity (Mus. Theobald and
Hanley).

10. **C. Ios,** Benson, An. Nat. Hist. ser. 2, vol. 10,
p. 350.—Pfeif. Mon. Helic. vol. 4, p. 761.
Darjiling.

PLATE XXV.

HELIX.

See previous plates, xiii. xiv. xv. xvi.

1. H. Chenui, var. Pfeiffer.

See also plate 27, f. 4. This form exhibits a close approach to the next species.

2. H. basilessa, Benson, An. Nat. Hist. (1865), ser. 3, vol. 5, p. 11.—Pfeif. Mon. Helic. vol. 5, p. 244. Travancore.

The worn example delineated (Benson's), has no epidermis. A perfect example has subsequently been obtained, which will be figured in a later plate.

3. H. Saturnia, Gould, Proc. Bost. Soc. N. H. vol. 2, p. 99.—Pfeif. Mon. Helic. vol. 3, p. 250; vol. 4, p. 300.
Tavoy.

4. H. ampulla, Benson, in Reeve, Conch. Icon. Helix. f. 736.—Pfeif. Mon. Helic. vol. 3, p. 27.
Nilgherries: Annamullay Hills.

5. H. cysis, Benson, in Pfeif. Mon. Helic. vol. 3, p. 92; vol. 4, p. 191.—H. cystis, Reeve, Conch, Icon. Helix, f. 737.
Nilgherries.

6. H. retrorsa, Gould, Boston. J. Nat. Hist. vol. 4, pl. 24, f. 5.—Pfeif. Mon. Helic. vol. 1, p. 76.
Birmah.

7. H. basileus, Benson, An. Nat. Hist. 1861, 1864.—Pfeif. Mon. Helic. vol. 5, p. 120.—H. Titanica, Pfeif. Proc. Zool. Soc. 1862, pl. 17, f. 3.
Annamullay Hills, S. India.

PLATE XXVI.

HELIX.

1. H. oxytes, Benson, J. Asi. Soc. Beng. 1836, vol. 5, p. 35.—Pfeif. Mon. Helic. vol. 1, p. 195.—Reeve, Conch. Icon. Helix, f. 734.
Bengal.

2, 5. H. Pollux, Theobald, J. Asi. Soc. Beng. 1859, vol. 27, p. 319: Notes Diag. pt. 2 (separate pamphlet), p. 7.—Var. as Nanina P. Blanford, J. Asi. Soc. Beng. 1870 (vol. 39, pt. 2), p. 13.
Near Teria Ghat : the var. from Nongkulong, &c., Khasi Hills.

3. H. Castor, Theobald, J. Asi. Soc. Beng. 1859, vol. 27, p. 319; Notes, Hist. pt. 2, p. 7.
Nunclai, Khasi Hills.

4. H. octhoplax, Benson, An. Nat. Hist. 1860, p. 190.—Pfeif. Mon. Helic. vol. 5, p. 400.
Western Skqo, Khasi Hills ; Moyang in Khasi, and near Asaloo in North Cachar.

6. H. Cherraensis, Blanford, J. Asi. Soc. Beng. 1870 (vol. 39, pt. 2), p. 14, pl. 3, f. 8 (as Nanina).
Cherra Punji, Khasi Hills.

Very closely allied, if, indeed, distinct from H. Castor.

7. H. cycloplax, Benson, An. Nat. Hist. s. 2, vol. 10, p. 34.—Reeve, Conch. Icon. Helix, f. 1156.—Pfeif. Mon. Helic. vol. 4, p. 181.
Darjiling, Sikkim Himalayah.

The keeled form (Benson's) here delineated runs into oxytes.

PLATE XXVII.

HELIX.

1, 2. H. Gordoniæ, Benson, An. Nat. Hist. 1863 (s. 3, vol. 11), p. 87.—Pfeif. Mon. Helic. vol. 5, p. 402.
Moulmein.

3. H. interrupta, Benson, Zool. Journ. vol. 5, p. 461 : Proc. Zool. Soc. 1834, p. 90.—Pfeif. Mon. Helic. vol. 1, p. 63; vol. 5, p. 122.—Reeve, Conch. Icon. Helix, f. 1159.
Sicrigalli, Bahar, and near the river Jellinghy.

4. H. Chenui, Pfeiffer, Mon. Helic. vol. 1, p. 438.—Reeve, Conch. Icon. Helic. f. 370.
Ceylon.

5. H. labiata, Pfeiffer, Proc. Zool. Soc. 1845, p. 65 : Mon. Helic. vol. 1, p. 73.
Landour.

6. H. thyreus, Benson, An. Nat. Hist. ser. 2, vol. 9 (1852), p. 405.—Pfeif. Mon. Helic. vol. 3, p. 251 ; vol. 4, p. 391.—Reeve, Conch. Icon. f. 735.
S. India.

Pfeiffer refers H. ryssolemma of Albers to this species.

7. H. Isabellina, Pfeif. Proc. Zool. Soc. 1851, p. 52 : Mon. Helix. vol. 4, p. 66.—Reeve, Conch. Icon. Helic. f. 1280.
Ceylon.

14 CONCHOLOGIA INDICA.

PLATE XXVIII.

HELIX.

1. **H. plicatula**, Blanford, J. Asi. Soc. Beng. 1870 (vol. 39, pt. 2), p. 13, pl. 17, f. 7 (as Nanina). Khasi Hills.

2. **H. Maderaspatana**, Gray, Proc. Zool. Soc. 1834, p. 67.—Pfeif. Mon. Helic. vol. 1, p. 63. Madras.

The H. Pondicherriensis of Pfeiffer, and the H. Korekonke of Férussac are regarded by Pfeiffer as identical.

3. **H. Haughtoni**, Benson, An. Nat. Hist. 1863 (ser. 3, vol. 11.), p. 87.—Pfeif. Mon. Helic. vol. 5, p. 92. Andamans.

4. **H. coraria**, Benson, An. Nat. Hist. s. 2. vol. 12, p. 91.—Pfeif. Mon. Helic. vol. 4, p. 67.—Reeve, Conch. Icon. Helic. f. 1291. Norton Plains, Ceylon.

5. **H. proxima**, Férussac, Hist. Moll. pl. 71, f. 5 — Pfeif. Mon. Helic. vol. 1, p. 377. Coimbatore, S. of Seringhapatam.

6. **H. solata**, Benson, An. Nat. Hist. 1848 (ser. 2, vol. 2), p. 159.—Pfeif. Mon. Helic. vol. 3, p. 67; vol. 4, p. 170.—Reeve, Conch. Icon. Helix, f. 741. Nilgherries.

According to Pfeiffer, Reeve's Menkeana is a form of this species.

7. **H. trochalia**, Benson, An. Nat. 1861. p. 82.—Pfeif. Mon. Helic. vol. 5, p. 329. Port Blair, Andamans.

8. **H. Orobia**, Benson, Au. Nat. Hist. 1848 (ser. 2, vol. 2), p. 158.—Reeve, Conch. Icon. Helix, f. 738. Himalayah.

9. **H. ligulata**, Férussac, Hist. Moll. pl. 31, f. 2.—Pfeif. Mon. Helic. vol. 1, p. 71.—Reeve, Conch. Icon. Helix, f. 395. Bundelkund, Bengal.

10. **H. cyclotrema**, Benson, An. Nat. Hist. 1863 (ser. 3, vol. 11), p. 88.—Pfeif. Mon. Helic. vol. 5, p. 123. Himalayah.

PLATE XXIX.

HELIX.

1. **H. bistrialis**, Beck, Index Moll. p. 2.—Pfeif. Mon. Helic. vol. 1, p. 71 (as = diaphana, Lea, and exilis, Chonn). — Reeve, Conch. Icon. Helix, f. 483. Tranquebar.

2. **H. Taprobanensis**, Dohrn, Malak. Blät. vol. 6 (1859), p. 206.—Pfeif. Mon. Helic. vol. 5, p. 116. Ceylon.

3. **H. Ceylanica**, Pfeiffer, Zeits. Malak. 1850, p. 67; Mon. Helic. vol. 3, p. 71; Kust. ed. Chemn. Hel. pl. 127, f. 6, 7.—Reeve, Conch. Icon. Helix. f. 1420. Ceylon.

4. **H. cyix**, Benson, An. Nat. Hist. ser. 3, vol. 5 (1860), p. 382.—Pfeif. Mon. Helic. vol. 5, p. 236. Matelle, Ceylon.

5. **H. Bombayana**, Grateloup, Act. Lin. Soc. Bordeaux, vol. 11, p. 406, pl. 1, f. 1.—Desh. in Férus. II. Moll. pl. 69, I. f. 5.—Reeve, Conch. Icon. Helix, f. 1194, and 1457 (as Belangeri).—Pfeif. Mon. Helic. vol. 1, p. 44, and vol. 3, p. 76. Bombay.

6. **H. Belangeri**, Deshayes, in Belang. Voy. Ind. Or. Zool. vol. 2, Mol. pl. 1, f. 1, 2, 3: Eneyc. Méth. Moll. vol. 2, p. 233.—Férus. Hist. Moll. vol. 1, p. 160, pl. 69, f. 4.—Pfeif. Mon. Helic. vol. 1, p. 69 (not Reeve). Pondicherry, Malabar.

7. **H. cestus**, Benson, J. Asi. Soc. Beng. vol. 5 (1836), p. 353 (amended in An. Nat. Hist. 1848, ser. 2, vol. 2, p. 182).—Reeve, Conch. Icon. Helix, f. 751.—Pfeif. Mon. Helic. vol. 3, p. 327. Bengal.

8, 9. **H. Helferi**, Benson, An. Nat. Hist. ser. 3, vol. 6, p. 191.—Pfeif. Mon. Helic. vol. 5, p. 366. Andamans.

10. **H. tugurium**, Benson, An. Nat. Hist. ser. 2, vol. 10, p. 348.—Reeve, Conch. Icon. Helix f. 1155.—Pfeif. Mon. Helic. vol. 4, p. 124. Darjiling.

PLATE XXX.

HELIX.

1. **H. anceps**, Gould, Boston, J. Nat. Hist. vol. 4, pl. 24, f. 6.—Pfeif. Mon. Helic. vol 1, p. 80.
Birmah.

2, 3. **H. hyba**, Benson, An. Nat. Hist. ser. 3, vol. 7 (1861), p. 83. — Pfeif. Mon. Helic. vol. 5, p. 181.
Dahinkboond, Sub-Himalayah, not far from the Sutlej.

4. **H. hemiopta**, Benson, An. Nat. Hist. 1863, ser. 3, vol. 11, p. 318.
Andamans.

5, 6. **H. palmaria**, Benson, An. Nat. Hist. ser. 3, vol. 13 (1864), p. 137.—Pfeif. Mon. Helic. vol. 5, p. 575.
Nundydroog, Mysore.

7. **H. basseinensis**, Blanford, J. Asi. Soc. Beng. 1865, p. 70 (Nanina).—Pfeif. Mon. Helic. vol. 5, p. 89.
Near Bassein, Aracan.

8, 9. **H. anlopis**, Benson, An. Nat. Hist. ser. 3, vol. 11 (1863), p. 318.—Pfeif. Mon. Helic. vol. 5, p. 93.
Port Blair, Andamans.

10. **H. propinqua**, Pfeiffer, Proc. Zool. Soc. 1857, p. 109.—Pfeif. Mon. Helic. vol. 4, p. 280.
Bombay.

PLATE XXXI.

HELIX.

1, 4. **H. bombax**, Benson, An. Nat. Hist. ser. 3, vol. 3 (1859), p. 186.—Pfeif. Mon. Helic. vol. 5, p. 151.
Moulmein.

The striæ are closer than can be represented in a lithograph: the mouth is not usually so broad.

2, 3. **H. Hodgsoni**, Blanford MSS. in Benson, An. Nat. Hist. ser. 3, vol. 3 (1859), p. 267.—Pfeif. Mon. Helic. vol. 5, p. 110.
Pankabari; Darjiling.

5, 6. **H. regulata**, Benson, An. Nat. Hist. ser. 3, vol. 5 (1860), p. 383.—Pfeif. Mon. Helic. vol. 5, p. 125.
Kahigauga and Katukellekande, Ceylon.

7, 10. **H. nuda**, Pfeiffer, Mon. Helic. vol. 3, p. 18.—Reeve, Conch. Icon. Helix, f. 781.
Himalayah.

8, 9. **H. politissima**, Pfeiffer, Proc. Zool. Soc. 1853, p. 125 : Mon. Helic. vol. 4, p. 45.—Reeve, Conch. Icon. Helix, f. 1292.
Ceylon.

PLATE XXXII.

HELIX.

1, 4. **H. chloroplax**, Benson, An. Nat. Hist. ser. 3, vol. 15 (1865), p. 14.—Pfeif. Mon. Helic. vol. 5, p. 80.
Near Simla, Himalayah.

2, 3. **H. Woodiana**, Pfeiffer (not Lea), Proc. Zool. Soc. 1851: Mon. Helic. vol. 3, p. 88 : Kuster ed. Chemn. Helix, pl. 114, f. 7, 8.—Reeve, Conch. Icon. Helix, f. 600.
Ceylon.

5, 6. **H. Noherensis**, Benson, An. Nat. Hist. ser. 3, vol. 13 (1864), p. 310.—Pfeif. Mon. Helic. vol. 5, p. 164.
Neher, Mahabaleshwar, W. India.

7, 10. **H. planiuscula**, Hutton, J. Asi. Soc. Beng. vol. 7, pt. 1 (1838), p. 218.—Pfeif. Mon. Helic. vol. 1, p. 60.
Simla; Landour.

8, 9. **H. molecula**, Benson, An. Nat. Hist. ser. 3, vol. 3 (1859), p. 389.—Pfeif. Mon. Helic. vol. 5, p. 69.
Rangoon, Birmah.

PLATE XXXIII.

CYCLOPHORUS.

See previous plates i. ii. iii. iv.

1, 7. **C. eximius**, Mousson, Moll. Java, p. 53, pl. 7, f. 1 (as Cyclostoma).—Reeve, Conch. Icon. Cyclop. f. 7.—Pfeif. Mon. Pneum. vol. 1, p. 69 : Kust. ed. Chemn. Cyclost. pl. 35, f. 1, 2 (as Cyclost.).
Khasia Hills.

2. **C. Ceylanicus**, Pfeiffer, in Kust. ed. Chenn.
p. 171, pl. 29, f. 1—3 (as Cyclostoma): Mon.
Pneum. vol. 1, p. 70. — Cyclostoma Indicum,
Sowerby, Thes. Conch. vol. 2, pl. 31 s, f. 320.—
Cycloph. Menkeanus, Reeve, Conch. Icon. Cycl.
f. 42, a, b.
Ceylon.

The figure 4 of *C. Indicus* in Belanger's work suits
this shell fairly, but neither 5 nor the description by
Deshayes will agree.

3. **C. Menkeanus**, Philippi, Zeitsch. Malak. 1847,
p. 123 (as Cyclostoma).—Reeve, Conch. Icon.
Cycloph. f. 42, c, d.—Pfeif. Mon. Pneum. vol. 1,
p. 66 : Kust. ed. Chenn. Cyclost. p. 171, pl. 28,
f. 6, 7, 8 (as Cyclostoma).
Near Point de Galle, Ceylon.

4. **C. aurantiacus**, Schumacher, Essai, p. 196 (for
Turbo volvulus, var. of Chenn. Conch. vol. 9,
f. 1061), as Annularia a.—Reeve, Conch. Icon.
Cycloph. f. 3.—Pfeif. in Kust. ed. Chenn. Cyclost.
pl. 4, f. 8, 9 (as Cyclostoma).
Moulmein Pagoda.

5, 6. **C. Jerdoni**, Benson, An. Nat. Hist. 1851 (ser. 2,
vol. 8), p. 185 (as Cyclostoma).—Pfeif. Mon.
Pneum. vol. 2, p. 51: Kust. ed. Chenn. Cyclost.
pl. 50, f. 1, 2, 3 (as Cyclostoma).
Nilgherries : Ceylon.

PLATE XXXIV.

CYCLOPHORUS.

1. **C. stenomphalus**, Pfeiffer, Zeitschr. Malak. 1846,
p. 44, and in Kust. ed. Chenn. Cyclost. pl. 8, f. 5, 6
(not his var. pl. 50), as Cyclostoma; Mon. Pneum.
vol. 1, p. 70 (not Reeve).
Khasia Hills.

2, 3. **C. altivagus**, Benson, An. Nat. Hist. ser. 2,
vol. 14 (1854), p. 411.—Pfeif. Mon. Pneum. vol. 2,
p. 57.—Reeve, Conch. Icon. Cycloph. f. 55.
Mahableshwar Hills, S. India.
The lower disk is devoid of either striae or ribs.

4. **C. Himalayanus**, Pfeiffer, Proc. Zool. Soc. 1851,
p. 242 (as Cyclostoma) : Mon. Pneum. vol. 1, p.
55 : Kust. ed. Chenn. Cyclost. pt. 2, p. 247 (not
fig.) as Cyclostoma H.—Reeve, Conch. Icon. Cycloph.
f. 11, a.
Darjiling, Sikkim Himalayah.
Approaches Aurora, but is broader (less so than

usual in the specimen delineated), is smoother below,
and with a painting composed on the spire of dense
angular brown lines.

5. **C. Bensoni**, Pfeiffer, Proc. Zool. Soc. 1852,
p. 158; Mon. Pneum. vol. 1, p. 63.—Reeve,
Conch. Icon. Cyclop. f. 38.
Northern part of Khasian Hills; Assam.
In the Annals of Natural History (1854, vol. 14,
p. 414), this was considered by Mr. Benson a varietal
form of C. Pearsoni ! It seems so of Reeve's.

6. **C. altivagus, var.** Benson.—Cyclostoma stenomphalus, var. Pfeif. in Kust. Chenn. pl. 50, f. 11—13
(not pl. 8, f. 5).
Bombay.

7. **C. sublaevigatus**, Blanford, Proc. Zool. Soc.
1862, p. 446.
Near Bhamo.
The type was kindly lent us by Mr. Blanford.

PLATE XXXV.

ACHATINA: Section Electra.

See previous plates, xvii. xviii.

1. **A. facula**, Benson, An. Nat. Hist. ser. 3, vol. 5
(1860), p. 466, for Perroteti, Reeve (not Pfeif.),
Conch. Icon. Achat. f. 102.—Pfeif. Mon. Helic.
vol. 6, p. 224.
Nilgherries.
The mouth is higher in proportion to its breadth
than is here represented.

2. **A. leptospira**, Benson, An. Nat. Hist. ser. 3,
vol. 15 (1865), p. 14.—Pfeif. Mon. Helic. vol. 6,
p. 233.
Soomeysur Hills.

3. **A. amentum**, Benson, in Reeve's Conch. Icon.
Achat. f. 82.—Pfeif. Mon. Helic. vol. 3, p. 499.
Lower Bengal, and Nerbudda.

4. **A. botellus**, Benson, An. Nat. Hist. ser. 3, vol. 5
(1860), p. 465.—Pfeif. Mon. Helic. vol. 6, p. 226.
Nilgherries.

5. **A. Vadalica**, Benson, An. Nat. Hist. 1865, Jan.
p. 15.—Pfeif, Mon. Helic. vol. 6, p. 229.
Wadalé, near Ahmednuggur.

The increase of the body-whorl above the mouth,
and the obliquity of the sutures, are inadequately
represented.

6. **A. Perrototi**, Pfeiffer, Mon. Helic. vol. 2, p. 260;
vol. 3, p. 491.—Glandina P. Philippi, Abbild.
N. C. vol. 1, Glandina, pl. 1, f. 12.—Var. A.
Nilagirica, Benson MSS. in Reeve, Conch. Icon.
Ach. f. 87.
Nilgherries.

7. **A. parabilis**, Benson, An. Nat. Hist. ser. 2,
vol. 18 (1856), p. 256.—Pfeif. Mon. Helic. vol. 4,
p. 606.
Ceylon.

8, 9. **A. notigena**, Benson, An. Nat. Hist. ser. 3,
vol. 5 (1860), p. 462.—Pfeif. Mon. Helic. vol. 6,
p. 429.
Mahaleshwar Hills, and near Bombay.

10. **A. Sarissa**, Benson, An. Nat. Hist. ser. 3, vol. 5
(1860), p. 463.—Pfeif. Mon. Helic. vol. 6, p. 234.
Lower Bengal.

PLATE XXXVI.

ACHATINA: Section Electra.

1. **A. crassilabris**, Benson, J. Asi. Soc. Beng.
vol. 5 (1836), p. 353.—Reeve, Conch. Icon. Achat.
f. 81.—Pfeif. Mon. Helic. vol. 2, p. 261; vol. 3,
p. 493.
Khasia Hills.

2. **A. panætha**, Benson, An. Nat. Hist. ser. 3, vol. 5
(1860), p. 384.—Pfeif. Mon. Helic. vol. 6,
p. 226.
Ceylon.

3. **A. Arthurii**, Benson, An. Nat. Hist. 1864, March,
p. 209.—Pfeif. Mon. Helic. vol. 6, p. 224.
Mahaleshwar Hills.

4. **A. crassula**, Benson, in Reeve's Conch. Icon.
Achat. f. 120.—Pfeif. Mon. Helic. vol. 3, p. 496.
Darjiling.

5. **A. Cassiaca**, Benson, in Reeve, Conch. Icon.
Achat. f. 86.—Pfeif. Mon. Helic. vol. 3, p. 439.
Khasia Hills.

6. **A. obtusa**, Blanford, Proc. Zool. Soc. 1869,
p. 449.
Bhamo, Upper Birmah.

The sub-genus Bacillum is proposed by Mr. Theobald
for this, the preceding, and other allied forms.

7. **A. gemma**, Benson, in Reeve's Conch. Icon. Achat.
f. 123 (and A. frumentum, ditto, f. 121).—Pfeif.
Mon. Helic. vol. 3, p. 496.
Lower Bengal, and Nerbudda.

8. **A. tenuispira**, Benson, J. Asi. Soc. Beng. 1836,
vol. 5, p. 353 (not vars. in An. Nat. Hist.).—Pfeif.
Mon. Helic. vol. 2, p. 262?
Khasia Hills: Darjiling.

9. **A. Shiplayi**; Pfeiffer, Mal. Blät. 1855, p. 169:
Mon. Helic. vol. 4, p. 612; Novit. vol. 1, pl. 22.
f. 13, 14.
Nilgherries.

10. **A. filosa**, Blanford, J. Asi. Soc. Beng. vol. 39.
pt. 2 (1870), p. 20, pl. 3, f. 18 (as Glessula, f.).
Travancore.

PLATE XXXVII.

STENOTHYRA and BYTHINIA.

1. **S. minima**, Sowerby, in Mag. Nat. Hist. (Charles-
worth's series), vol. 1 (1837), p. 217, f. 22, b (as
Nematura).—Adams, Proc. Zool. Soc. 1851 (as
Nematura).—Benson, An. Nat. Hist. ser. 2, vol. 17
(1856), p. 501.
Western India.

2. **S. Deltæ**, Benson, J. Asi. Soc. Beng. vol. 5 (1836).
p. 78 (as Nematura); An. Nat. Hist. ser. 2, vol. 17
(1856), p. 499.—Sow. Mag. Nat. Hist. (Charles-
worth's series), vol. 1 (1837), p. 217 (as Nematura).
River Hooghly, near Calcutta; also in a salt
lake near Ballinghat.

3. **S. foveolata**, Benson, An. Nat. Hist. ser. 2.
vol. 17, p. 497.
Ganges, near Sikrigali.

Delineated from the worn type, the only example
known to us.

4. **S. monilifera**, Benson, An. Nat. Hist. ser. 2,
vol. 17 (1856), p. 497.
Mergui, Birmah.

5, 6. **B. inconspicua**, Dohrn, Proc. Zool. Soc. 1857.
p. 123.
Ceylon.

7. **B. lutea**, Gray, Annals Phil. 1824, p. 277.—
B. gonioatoma, Hutton MSS.—Paludina pulchella,
Kust. (not Bens.) ed. Chem. Palud. p. 30, pl. 6,
f. 19.
Purneah.

8, 9. **B. Nassa**, Theobald, J. Asi. Soc. Beng. vol. 34,
pt. 2 (1865).
Shan States.

10. **B. Iravadica**, Blanford, Proc. Zool. Soc. 1869,
p. 446.
Marshes near Mandelay, Birmah.

PLATE XXXVIII.

BYTHINIA.

1, 4. **B. Cerameopoma**, Benson, Gleanings Sc.
Calcut. vol. 2, p. 125 (name for species in vol. 1,
p. 563), as Paludina: J. Asi. Soc. Beng. vol. 24
(1855), p. 131.—B. Ceranospatana, Frauenfeld,
Verhandl. Zool. Bot. Wien. 1862, p. 1156.
River Ken : Coosa River.

2, 3. **B. Travancorica**, Benson, An. Nat. Hist.
ser. 3, vol. 6 (1860).
Near Quilon, Travancore.
The minute spiral striolæ are characteristic.

5, 6. **B. pulchella**, Benson, J. Asi. Soc. Beng. vol. 5
(1836), p. 746.
Sylhet.
This is stated to be the Valvata no. 9 of Hutton in
the third volume of the Journal of the Asiatic Society
of Bengal.

7, 10. **B. stenothyroides**, Dohrn, Proc. Zool. Soc.
1857, p. 123.
Ceylon; Nilgherries; Poonah; Madras;
S. Arcot; Trichinopoli, &c.

8, 9. **B. orcula**, Benson, MSS. in Frauenfeld,
Verhandl. Zool. Bot. Wien. 1862, p. 1154.
Purneah.

PLATE XXXIX.

PLANORBIS.

1, 2, 3. **P. calathus**, Benson, An. Nat. Hist. ser. 2,
vol. 5 (1850), p. 349.
Near Moradabad; Kattiawar; Ceylon ; Cashmire.

4, 5, 6. **P. Trochoideus**, Benson, J. Asi. Soc. Beng.
vol. 5 (1836), p. 742 (Glean. Scien. Calcutta, vol.
1, pl. 8, f. 10); An. Nat. Hist. (1850), ser. 2, vol.
5, p. 352.
Barrakpore.

7, 8, 9. **P. coenosus**, Benson, An. Nat. Hist. 1850,
p. 349.
Near Moradabad : Ceylon.

10. **P. exustus**, Desh. Belang. Voy. Ind. Orient. Zool.
p. 417, pl. 1, f. 11-13 (1834): ed. Lam. vol. 8, p.
392.—Müller, Synopsis Test. p. 34. P. Indicus,
Benson, J. Asi. Soc. Beng. vol. 5, (1836), p. 743.
—Martens, Mal. Blätt. vol. 14, p. 212.
Universally diffused, from Bombay to Cashmire.
Otho Müller (Verm. Ter. et. Fluv. pt. 2, p. 157),
indicated this shell as the Coromandel form of corneus;
hence probably the manuscript name of P. Coroman-
delianus (Kuster, as of O. Fabricius). Chemnitz
figured it (Conch. Cab. vol. 9, pl. 127, f. 116, 117), as
the Indian variety of the same species, to which it was
referred with doubt, also, by Hutton (J. Asi. Soc. vol.
3). A figure of it appeared in the Gleanings of Science
(vol. 1, pl. 8, f. 6).

PLATE XL.

PLANORBIS.

1, 2, 3. **P. Cantori**, Benson, An. Nat. Hist. 1850,
p. 349.
Barrakpore.

4, 5, 6. **P. Sindicus**, Benson, An. Nat. Hist. (1850),
p. 350.
River Indus, Upper Sinde.
Figured from the type, the only example known
to us.

7, 8, 9. **P. umbilicalis**, Benson, J. Asi. Soc. Beng.
vol. 5 (1836), p. 741.—An. Nat. Hist. ser. 2, vol. 5
(1850), p. 341. — Martens, Mal. Blätt. vol. 14,
p. 216.
E. Bengal.

10. **P. exustus**, Deshayes.—See plate 39, f. 10.
The species is often much flatter and larger than the
specimen delineated.
A form with a very contracted aperture was taken
from tanks near Chanda.

PLATE XLI.

UNIO.

See previous plates, ix. to xii.

1. **U. Layardi**, Lea, Proc. Ac. Nat. Sc. Philadelph. 1850, p. 155 : Journ. Ac. Nat. Sc. Philad. ser. 2, vol. 4 (and Obser. U. vol. 7), pl. 56, f. 122.— Reeve, Conch. Icon. Unio, f. 111.

Ceylon.

We are unable to distinguish the exact line of demarcation between this and Thwaitesii; in characteristic specimens, however, the front extremity is longer and more tapering.

2. **U. involutus**, Benson, in Hanley's Recent Bivalves, p. 385, pl. 23, f. 19.—Reeve, Conch. Icon. Unio, f. 177.

Assam.

3. **U. favidens**, var. chrysis, Benson, Ann. Nat. Hist. ser. 3, vol. 10 (1862), p. 188.

River Dojum at Kareily Ghát, near Bareilly.

4. **U. Jenkinsianus**, Benson, An. Nat. Hist. ser. 3, vol. 10 (1862), p. 185.

River Berhampooter, Assam.

The unique type here figured will probably be considered an abnormal form of marginalis or Corrianus.

5, 6. **U. Nuttallianus**, Lea, Journ. Ac. Nat. Sc. Philad. ser. 2, vol. 3, p. 310, pl. 30, f. 23 : Obser. U. vol. 6, p. 30, pl. 30, f. 23.

Assam.

7. **U. consobrinus**, Lea, Journ. Ac. Nat. Sc. Philad. ser. 2, vol. 4 (and Obser. U. vol. 7), pl. 90, f. 122. —Benson, An. Nat. Hist. ser. 3, vol. 10 (1862), p. 193.

Cochin, Malabar.

Only two specimens, both in the Bensonian collection, are known to us. They approach marginalis, of which they may possibly prove a tumid variety, yet have a fibrous, not a satin-like, style of epidermis, a different shape, and a different disposition of colouring, the yellow band not being adjacent to the ventral edge. There is no appearance, moreover, of the two raised lines which adorn the umbonal slope near the beaks, which, unfortunately being eroded, do not exhibit any marked character. The front extremity is more prominent and rounded (not obliquely cut off below) than in the allied forms. According to Lea the original types came from China.

PLATE XLII.

UNIO.

1. **U. Birmanus**, Blanford, Proc. Zool. Soc. 1869, p. 450.

Bhamo, Upper Birmah.

2. **U. favidens**, var. Deltæ, Benson, An. Nat. Hist. ser. 3, vol. 10 (1862), p. 188.

River Jellinghy, Upper Gangetic Delta, Bengal.

The radiation spoken of consists of almost imperceptible lines. A specimen from Tirhoot is of an uniform dark olivaceous hue.

3. **U. foliaceus**, Gould, Proc. Boston Soc. N. H. vol. 1, p. 141 : Otia Conch. p. 191.—U. Peguensis, Anthony, American J. of Conch. vol. 1 (1865), p. 351, pl. 23, f. 2.—Reeve, Conch. Icon. Unio, f. 519.

Tavoy, Birmah ; Pegu.

Gould's type (sent to Benson), is unmistakably the young of the adult Peguensis.

4, 5, 6. **U. marcens**, Hanley, for U. favidens, var. marcens of Benson, An. Nat. Hist. ser. 3, vol. 10 (1862), p. 188.

Berhampooter River, Assam.

The absence of all corrugation is a most important characteristic feature.

7. **U. marginalis**, var. Anodontina.—U. Anodontinus, Kuster (not Lamarck), ed. Chemn. Conch. Unio, pl. 80, f. 5.

River Godavery ; Nagpoor ; Sylhet.

Although we have not fully traced the intervening links between this elongated form and the typical marginalis, we cannot doubt its approximation to those specimens which have been supposed (perhaps erroneously) to represent the U. lälineatus of Lea. Virginia is the recorded, and probably the correct, locality of the Lamarckian U. Anodontoides, but the types are declared to be Indian, not American; its dentition is said to be inconspicuous. In the specimen delineated the front teeth are short, strong, and very oblique; the nacre is of a rather pale salmon-colour.

PLATE XLIII.

UNIO.

1. **U. Thwaitesii**, Lea, Proc. Philad. Ac. N.S. vol. 1 (1859), p. 152; J. Philad. s. 2, vol. 4, pl. 37, f. 125.—Reeve, Conch. Icon. Unio, f. 105. Ceylon.

U. marginalis, Lamarck, Anim. s. Vert. ed. Desh. vol. 6, p. 541.—Hanley, Rec. Bivalves, p. 206, pl. 19, f. 53.—U. testudinarius, Spengler, Skriv. Nat. Selks. vol. 3, pl. 1, p. 65, and U. truncatus, p. 56 (fide March).—U. Grœnlandicus (in Lea) from Schröter, Fluss. Conch. p. 181, pl. 9, f. 1.

We describe the hinge of the typical form, which is but little modified in any of the varieties. In the right valve are two sloping anterior teeth, of which the upper and narrower arches more or less downwards, and the lower is strong and rather large; the single lateral one is strong, elongated, and bent at the end. In the left valve, besides the callus, there is one anterior and two lateral teeth, the latter scarcely divided until the second moiety, where they slant down. The principal anterior scar is decidedly large in proportion to its size in the allied species. After long examination of many scores of examples of this most variable species from nearly every part of British India, we can only arrive at the conclusion that the forms usually designated in cabinets, bilineatus, lamellatus, Corrianus, &c., all run into each other. We do not assert, however, that the shells intended by Lea are identical; for his figure of bilineatus looks like the young of some broadly-winged Siamese shell (of the Housei type), and his lamellatus is notable for a peculiarity of dentition. We have delineated some of the more striking forms.

2. **U. marginalis**, var. typica.—U. marginalis, Encycl. Méth. Vers, pl. 247, f. 1. Moulalabad : Pegu.

The brief Lamarckian diagnosis is further defined by his reference, &c., and his figure.

3. **U. marginalis**, var. obesa. River Irawadi, Birmah.

A giant form, which does not exhibit the ochraceous band, and is peculiarly swollen. It comes between the var. lata, and the typical form. The upper anterior tooth is almost linear; the lateral are not bent at the extremity, and the upper one in the left valve is scarcely developed.

4. **U. marginalis**, var. Candaharica, Hutton, J. Asi. Soc. Beng. vol. 17, pt. 1 (1849), p. 651. River Sutlej.

The abnormal characters of shape and colouring are such that Hutton (its discoverer) suggested its possible distinctiveness. The nucleus very closely approaches U. them, Benson; the adult cannot be separated from the forto bilineatus.

5. **U. marginalis**, var. tricolor.—U. tricolor, Kuster. ed. Chenn. Unio, pl. 45, f. 1?

A very beautifully painted shell, which is usually more or less compressed.

PLATE XLIV.

UNIO.

1. **U. marginalis**, var. cylindrica.

2. **U. marginalis**, var. zonata.—U. marginalis, Desh. Encycl. Méth. Vers, vol. 2, p. 587. Belgaum, Deccan.

3. **U. marginalis**, var. bilineata.—U. bilineatus, Reeve (as of Lea), Conch. Icon. Unio, f. 365.

Nearly all the varieties (especially the immature examples) exhibit two raised lines near the beaks on the umbonal slope. We are aware that Spengler intended to have indicated U. marginalis as U. testudinarius, but his Latin description is most utterly inadequate to define it; his delphinus, conus, and gibbosus (from the East Indies), must be ignored for a similar reason. The name Grœnlandicus (the locality was subsequently corrected in the Einleitung, vol. 2, p. 621, by Schröter himself), would mislead.

4. **U. marginalis**, var. Corriana.—U. Corrianus, Lea, Trans. Amer. Phil. Soc. vol. 5, p. 177, pl. 9, f. 25, from which Hanley, Rec. Bivalves, p. 207, pl. 21, f. 60. Near Calcutta, &c.

This form, always more or less indented in the middle, is sometimes elongated cylindrical, sometimes compressed and oval-oblong. The cardinal callosity (rarely absent) is more or less developed, and the anterior or hinge teeth are almost horizontal, curve outward, and are either rather elongated, or if shorter, are rather prominently elevated.

5, 6. **U. corrugatus**, var. levirostris.—U. levirostris, Benson, An. Nat. Hist. ser. 3, vol. 10 (1862), p. 191 (from types).

River Godavery; Pemgunga, &c.

Benson's original examples were much worn at the beaks, hence the name. A perusal of Schröter's description of Mya corrugata, in his Flussconchylien, shows that he was perfectly aware of the frequent absence of the characteristic corrugation.

7. **U. lamellatus**, Reeve (as of Lea, Trans. Amer. Phil. Soc. s. 2, vol. 6, and Obs. Un. vol. 2, p. 19, pl. 6, f. 16, from which Hanl. Rec. Biv. Shells, p. 194, pl. 21, f. 50), Conch. Icon. Unio, f. 511.

Pegu.

Closely allied to U. generosus. The teeth, however, are peculiarly elongated. The form delineated is abnormally high; other examples since obtained are much narrower, and more like Lea's figures. The long lamellar hinge-teeth referred to by Lea, are very manifest, but whether the species is distinct from generosus may be doubted. Young examples are olivaceous yellow, changing to dark green on the very concave posterior slope.

PLATE XLV.

UNIO.

1. **U. crispatus**, Gould, Proc. Boston Soc. N. Hist. vol. 1, p. 141; Otia Conch. p. 191.

Tavoy, Birmah.

The specimen figured was sent to Benson by the American describer.

2 to 5. **U. corrugatus**, Müller.—Mya c. Müller, Beschaft. Ges. Naturf. Berlin, vol. 4, p. 58, pl. 3, f. 7.—Chemn. Conch. Cab. vol. 6, p. 31, f. 22 (from which Kust. ed. Chemn. Unio, pl. 97, f. 3, 4).—Mawe, Lin. Conch. pl. 4, f. 3.—Not of Martini and Wood (as Mya) or Reeve (as Unio).—Mya spuria, Gmel. Syst. 3222, from Schröter, Einleit. pl. 7, f. 5 (copied as U. concentricus, Valenc. in Enc. Méth. pl. 249, f. 3).

Near Madras; River Godavery; Nagpore; Pemgunga, &c. &c.

Authors have recognised very different shells as the fragile and pellucid Mya corrugata of Müller (Verm. pt. 2, p. 211). His original description being utterly insufficient, had better be ignored; he defined the species, however, by his figure in the Berlin journal. We know of no adequate representations in the older works, hence Wood supposed it to be the very coarsely sculptured Cingalese (?) species, which we have called Tennentii, whilst Benson and others thought it faviolens. The views of the exterior given by Chemnitz and Mawe are indefinite, the outlines of the interior (and the hinge in Mawe's figure) suit the present species. The range of character from entire smoothness (except near the tips) to a coarse divaricate corrugation over the dorsal half of the surface, from tumidity to compression, from thinness to solidity, from olive green to ochraceous green, can only be rivalled by its diversity of contour; every link, however, has been most cautiously traced. We believe that the U. Nagporensis of Lea (Journ. Ac. Nat. Sc. Philad. ser. 2, vol. 1, and Obs. U. vol. 7, p. 88, pl. 15, f. 150), will also prove a large variety, with worn beaks, of our form fig. 3, but dare not assert so positively.

2. Var. solida, from the River Godavery.

3. Var. Nagporensis. — ? U. Nagporensis, Lea, Journ. Ac. Nat. Sc. Philad. ser. 2, vol. 4 (Obs. Un. vol. 7, p. 88), pl. 15, f. 150.

Nagpore; Pemgunga.

The coincidence of locality and the general contour render it probable that Lea founded his species upon a worn aged example of this swollen form.

4. Var. fragilis.

A thin ventricose form, which, we are assured, has been taken from the inside of fishes.

5. **U. corrugatus.** Typical form from Madras and Southern India.

6. **U. Wynegungaensis**, Lea, Proc. Ac. Nat. Sc. Philad. vol. 8, p. 331 (Obs. Un. vol. 7, p. 89), pl. 15, f. 151. Reeve, Conch. Icon. Unio, f. 539.

River Wynegunga. &c.

7, 8, 9. **U. Tennentii**, Hanley, for Mya corrugata, Wood, General Conch. p. 108, pl. 24, f. 1, 2, 3.
" Ceylon."

Our locality is not well authenticated. The shell was found without name in the Bensonian collection, as from British India.

10. **U. corbis**, Benson, in Hanley's Recent Bivalves, p. 386, pl. 23, f. 43.
Assam.

So rare a shell, that we have only seen one perfect specimen. It is a solid little species.

PLATE XLVI.

UNIO.

1. **U. scutum**, Benson's MSS. in Reeve's Conch. Icon. Unio, f. 510.
Tenasserim River.

2. **U. scobina**, Benson, in Hanley's Recent Bivalves, p. 382, pl. 22, f. 40.
Assam; Mysore?

The specimen delineated is the almost unique original type, cleared partially of the ferruginous coating with which it was invested. The Mysore specimens are too worn to be positively pronounced identical.

3. **U. scobina?** var. Benson.
Belgaum, Deccan.

A half link between scobina and ovatus.

4. **U. generosus**, Gould, Proc. Boston Soc. Nat. H. vol. 2, p. 220; Otia Conch. p. 201.

The shell here delineated was sent by Gould to Benson as typical; the alated form referred by us to lamellatus (pl. 9) belongs to it. Except for its dentition, this shell might be referred to the large Birmese form of marginalis, or even to the Cingalese Thwaitesii. Its affinities, on the other hand, incline closely to the U. lamellatus of Lea. The peculiar hinder surface, which Gould terms subcostated, results from a close superficial concentric sulcation.

5, 6. **U. Bonneaudi**, var. Eydoux and Souleyet.—Reeve, Conch. Icon. Unio, f. 515.
Pegu.

The sculpture is scarcely so strong in the smooth variety delineated as in another of our figures, and the shading here obscures the radiating sulci in front.

7. **U. generosus**, Gould, var. angustior.
Pegu.

PLATE XLVII.

CYCLOPHORUS and AULOPOMA.

See previous plates i. ii. iii. iv. xxxiii. xxxiv.

1, 2. **A. grande**, Pfeiffer, Proc. Zool. Soc. 1855, p. 104: Mon. Pneum. vol. 2, p. 39: Novit. Conch. vol. 1, pl. 19, f. 11 to 13.
Ceylon.

3, 4. **A. Hoffmeisteri**, Troschel, Zeitsch. Malak. 1847, p. 43.
Ceylon.

The description referred to is only in German, and very brief.

5, 6. **C. hispidulus**, Blanford, Jour. Asi. Soc. Beng. vol. 32 (1863), p. 324; Cont. Ind. Mal. pt. 4, p. 3.
Mya Leit Doung, near Ava.

The spiral hispid ridges of the umbilicus are not adequately rendered in our lithograph.

7. **C. exul**, Benson, An. Nat. Hist. ser. 2, vol. 11 (1854), p. 412.—Pfeif. Mon. Pneum. vol. 2, p. 46.—Reeve, Conch. Icon. Cyclop. f. 53.
Bhamoury, at the foot of the Kobillano-Himalayahs.

The delicate shell here depicted is unique.

8. **C. cratera**, Benson, An. Nat. Hist. ser. 2, vol. 18 (Aug. 1856).—Pfeif. Mon. Pneum. vol. 2, p. 55.
Ceylon.

Possibly a form of the annulatus of Troschel.

9. **C. cytopoma**, Benson, An. Nat. Hist. ser. 3, vol. 5 (1860), p. 385.—Pfeif. Mon. Pneum. vol. 3, p. 72.
Ceylon.

Scarcely differs from the previous species, except in fragility and the want of a double lip.

10. **C. tryblium**, Benson, var. conica. See plate xlviii. f. 1.

The typical form is delineated in our next plate: that here represented is remarkable for its narrow umbilicus and its elevated spire.

PLATE XLVIII.
CYCLOPHORUS.

1. **C. tryblium**, Benson, Ann. Nat. Hist. ser. 2, vol. 14 (1854), p. 412.—Pfeif. Mon. Pneum. vol. 2, p. 45.

Darjiling; Sikkim Himalayah.

2. **C. affinis**, Theobald. See pl. 2, f. 7.

The peculiarity of lip referred to by the author was simply accidental. The individual here represented, which reminds one of a dwarf carinated Siamensis, is almost the only one known to us, the supposed second example alluded to having proved to be distinct.

3. **C. Indicus**, Pfeiffer (as of Deshayes), Mon. Pneum. vol. I, p. 77 (from Cuming's collection); Kuster, ed. Chemn. Cyclos. pl. 33, f. 3, 4 (as Cyclostoma).

Isle of Elephanta, near Bombay.

It may be doubted whether this can be the Cyclostoma Indicum of Deshayes in Belanger's Voyage (Zool. Moll. pl. 1, f. 4, 5), the lower face of which does not suit any example we have seen.

4. **C. Malayanus**, Benson, An. Nat. Hist. ser. 2, vol. 10 (1852), p. 269 (as Cyclostoma).—Pfeif. Mon. Pneum. vol. 2, p. 42.—Reeve, Conch. Icon. Cycloph. f. 2.

Shan States.

The previously recorded locality was Pulo Penang. The Cyclostoma volvulus of Soukyet (Voy. Bonite, Moll. pl. 30, f. 18-21) is considered by Pfeiffer to be identical.

5. **C. Pearsoni**, Benson. See plate 1, f. 6.

We have figured Benson's type, which displays no traces of that articulated keel which forms so prominent a feature in the next species.

6. **C. Haughtoni**, Theobald. See plate 1, f. 3, and plate 3, f. 6.

A characteristic form. In our previous figures the articulation was inadequately represented, few, if any, traces of it having been visible upon the portion delineated, though freely developed upon the opposite side.

7. **C. Siamensis**, Sowerby, Thesaur. Conch. vol. 1, p. 158*, pl. 31, f. 392, 3 (as Cyclostoma).—Benson, An. Nat. H. ser. 2, vol. 19.—Reeve, Conch. Icon. Cycloph. f. 19.—Pf. Mon. Pneum. vol. 1, p. 56.

Lacat, and Teria Ghat.

PLATE XLIX.
PTEROCYCLOS, including SPIRACULUM, and RHIOSTOMA.

See previous plate v.

1, 2. **P.** (Sp.) **Fairbanki**, Blanford, J. Asi. Soc. Beng. 1869, p. 135, and Cont. Ind. Mal. pt. 10 (as Spiraculum).

Pulney Hills, S. India.

3, 4. **P.** (Sp.) **Andersoni**, Blanford, Proc. Zool. Soc. 1869, p. 447 (as Spiraculum).

Near Bhamo, Upper Birmah.

5, 6. **P. nanus**, Benson, An. Nat. Hist. ser. 2, vol. 8 (1851), p. 456.—Reeve, Conch. Icon. Pter. f. 12.

Nilgherries.

7, 8. **P. Cumingi**, Pfeif. Proc. Zool. Soc. 1852, p. 158; Mon. Pneum. vol. 2, p. 29.—Reeve, Conch. Icon. Pter. f. 14.

Ceylon.

9, 10. **P.** (Sp.) **Gordoni**, Benson, An. Nat. Hist. 1863, May (as Opisth.).—Pfeif. Mon. Pneum. vol. 3, p. 36 (as Pt.).—Theob. J. Asi. Soc. Beng. 1870, vol. 39, pt. 2, p. 399, pl. 18, f. 6.

Sittoung, near Tonghu; near Moulmein. Birmah.

PLATE L.
HELIX.

1, 2. **H. mammillaris**, Blanford, J. Asi. Soc. Beng. 1865, p. 69, and Cont. Mal. pt. 5 (as Nanina). Pfeif. Mon. Helic. vol. 5, p. 88.

Akoutong, Pegu.

3, 4. **H. Helicifera**, Blanford, J. Asi. Soc. Beng. 1866, p. 34, and Cont. Mal. pt. 5 (as Nanina).

Aracan Hills, near Prome.

5. **H. acuducta**, Benson, An. Nat. Hist. ser. 2, vol. 5 (1850), p. 214.—Reeve, Conch. Icon. Helix. f. 739.—Pfeif. Mon. Helic. vol. 3, p. 78; vol. 4, p. 67.

Nilgherries.

6. **H. Travancorica**, Benson, An. Nat. Hist. (1865), ser. 3, vol. 15, p. 13.—Pfeif. Mon. Helic. vol. 5, p. 131.
Hills of Travancore.

Lithography cannot adequately represent the rasp-like decussation of the sculpture of this species. The whorls are concave above the very sharp keel.

7. **H. serrula**, Benson, J. As. Soc. Beng. vol. 5 (1836).—Pfeif. Mon. Helic. vol. 1, p. 206.
Bengal.

8, 9. **H. Austeni**, Blanford, J. As. Soc. Beng. 1870 (vol. 39, pt. 2), p. 15, pl. 3, f. 10, and Cont. Ind. Mal. pt. 11 (as Nanina).
Habiang, in Garo Hills.

10. **H. crassicostata**, Benson, An. Nat. Hist. 1848 (ser. 2, vol. 1), p. 159.—Reeve. Conch. Icon. Helix, f. 747.
Southern India.

PLATE LI.

HELIX : Section Zonites.

1. **H. choinix**, Benson, An. Nat. Hist. 1864 (ser. 3, vol. 7), p. 185.—Pfeif. Mon. Helic. vol. 5, p. 117.
Andaman Isles.

2. **H. acerra**, Benson, An. Nat. Hist. ser. 3, vol. 3 (1859).—Pfeif. Mon. Helic. vol. 5, p. 100.
Mergui, Tenasserim.

3. **H. pedina**, Benson, An. Nat. Hist. ser. 3, vol. 15 (1865), p. 13.—Pfeif. Mon. Helic. vol. 5, p. 100.
Near Bombay, and Ahmednugger.

4. **H. resplendens**, Philippi, Zeits. Malak. 1846, p. 192.—Pfeif. Mon. Helic. vol. 5, p. 56; vol. 5, p. 100.—H. expolita, Desh. teste Pfeif.
Birmah.

5, 6. **H. rubellocincta**, Blanford, J. As. Soc. Beng. 1870, vol. 39, pt. 2, p. 14, pl. 3, f. 9, and Cont. Ind. Mal. pt. 11 (as Nanina).
Habiang in Garo Hills, beyond the southern boundaries of Assam.

7, 10. **H. splendens**, Hutton, J. As. Soc. Beng. vol. 7, pt. 1, p. 215 (as Nanina).—Pfeif. Mon. Helic. vol. 1, p. 73; vol. 4, p. 124.
Mahassu, Fagu, &c., Himalayah.

8, 9. **H. lubrica**, Benson, An. Nat. Hist. ser. 2, vol. 10 (1852), p. 349.—Pfeif. Mon. Helic. vol. 4, p. 44.—Reeve, Conch. Icon. Helix, f. 1153.
Himalayah.

PLATE LII.

HELIX.

1. **H. Nicobarica**, Beck, Index Moll. Crist. Frcd. Ap. p. 5 (as Nanina).—Pfeif. Mon. Helic. vol. 1, p. 40.—Reeve, Conch. Icon. Helix, f. 1157.
Cuddapah.

This rare shell was figured by Chemnitz (vol. 9, f. 911, 912), as a sinistral Nicobar form of H. ponnatia.

2, 5. **H. textrina**, Benson, An. Nat. Hist. ser. 2. vol. 17 (1856), p. 252.—Pfeif. Mon. Helic. vol. 4, p. 58; Novit. vol. 1, pl. 36, f. 5 to 7.—Martens. Ostas. Zool. vol. 2, p. 71 (as Nanina).
Tongoop Hills, Aracan; Birmah.

3. **H. monticola**, Hutton, Jour. As. Soc. Beng vol. 7, pt. 1, p. 215 (as Nanina).—Pfeif. Mon. Helic. vol. 1, p. 130 (not vol. 5, p. 197).
Huttu and Mahassu, Himalayah.

4. **H. basilessa**, Benson. See plate xxv. f. 2.

6. **H. albixonata** (var.), Dohrn, Proc. Zool. Soc. 1858, p. 133; Pfeif, Mon. Helic. vol. 5, p. 90.
Trichoor (Benares); Ceylon.

7. **H. Juliana** (var. Reevei), Gray, Proc. Zool. Soc. 1834, p. 58 (as Nanina).—Reeve, Conch. Icon. Helix, f. 373.
Ceylon.

A very beautiful form, but not much like that originally described. As other varieties will be figured, no synonymy is here appended.

PLATE LIII.

HELIX.

1. **H. similaris**, Férussac, Hist. Moll. pl. 25, B. f. 1-4.—Pfeif. Mon. Helic. vol. 1, p. 336.—Reeve, Conch. Icon. Helix, f. 767.—H. Woodiana, Lea, and translucens, King, teste Pfeif.
Pegu.

2. **H. similaris** (var.) Férussac.—Allied to below and Peguensis.

3, 4. **H. scenoma**, var. Benson.

5. **H. scenoma**, Benson, An. Nat. Hist. 1863 (ser. 3, vol. 11), p. 319.—Pfeif. Mon. Helic. vol. 5, p. 331.
Near Moulmein.

6. **H. pilidion**, Benson, An. Nat. H. 1860 (ser. 3, vol. 6), p. 191.—Pfeif. Mon. Helic. vol. 5, p. 347.
Pegu.

7. **H. bolus**, Benson, An. Nat. Hist. ser. 2, vol. 18 (1856), p. 252.—Pfeif. Mon. Helic. vol. 1, p. 251.
Thyet Mio.

8. **H. hemiopta**, Benson, An. Nat. Hist. 1863 (s. 3, vol. 11), p. 318. See also pl. xxx. f. 4.
Andaman Isles.

9. **H. scalpturita**, Benson, An. Nat. Hist. 1859 (ser. 3, vol. 3), p. 394.- Pfeif. Mon. Helic. vol. 5, p. 349.
Ava

10. **H. coriaria** (var.), Pfeiffer, Zeits. Malak. 1847, p. 145.—Reeve, Conch. Icon. Helix, f. 417.
Ceylon.

Australian specimens cannot be distinguished from the shell here delineated.

PLATE LIV.

HELIX.

1, 2. **H. Cingalensis**, Benson, An. Nat. Hist. 1863 (ser. 3, vol. 5), p. 185.—Pfeif. Mon. Helic. vol. 5, p. 93.—H. Emiliana, Reeve (not Pfeif.), Conch. Icon. Helix, f. 608.
Matelle, Ceylon.

3. **H. hyphasma**, Pfeiffer, Proc. Zool. Soc. 1855 p. 124; Mon. Helic. vol. 4, p. 40.—Reeve, Conch. Icon. Helix, f. 1297.
Ceylon.

4. **H. cacuminifera**, Benson, An. Nat. Hist. 1850 ser. 2, vol. 5, p. 214.—Pfeif. Mon. Helic. vol. 4, p. 36.—Reeve, Conch. Icon. Helix, f. 744.
Nilgherries.

5. **H. apicata**, Blanford (as Nanina, section Trochomorpha), J. Asl. Soc. Beng. 1870 (vol. 39, pt. 2). p. 16, pl. 3, f. 13; Cont. Ind. Mal. pl. 11, p. 16.
Nilgherries in Southern India, at Coonoor. Neddiwuttum, &c.

6. **H. aeris**, Benson, An. Nat. Hist. 1859 (ser. 3 vol. 3), p. 387.—Pfeif. Mon. Helic. vol. 5, p. 355.
Teria Ghat; Khasia Hills.

7. **H. galea**, Benson, An. Nat. Hist. 1859 (ser. 3. vol. 3), p. 388.--Pfeif. Mon. Helic. vol. 5, p. 264.
Teria Ghat; Birmah.

8. **H. arx**, Benson, An. Nat. Hist. 1859 (ser. 3. vol. 3), p. 384.—Pfeif. Mon. Helic. vol. 5, p. 98.
Therabuin Hills, Tenasserim Valley.

9. **H. infula**, Benson, An. Nat. Hist. 1848 (ser. 2. vol. 2), p. 160 (changed from turbiniformis, Cabot, J. Nat. Hist.).—Reeve, Conch. Icon. Helix, f. 785.
Rajmahal Hills.

10. **H. falcata**, Blanford, J. Asi. Soc. Beng. 1870 vol. 39, pt. 2, p. 15, pl. 3, f. 11, and Cont. Ind. Mal. p. 15 (as Nanina).
Haldang in Garo Hills, beyond Southern boundaries of Assam.

PLATE LV.

HELIX.

1. **H. Guerini**, Pfeiffer, Revue Zool. Soc. Cuv. 1842 p. 304; Mon. Helic. vol. 1, p. 118.—Philip. Abbild. N. Conch. vol. 1, pl. 3, f. 6.
Nilgherries

2. **H. camura**, Benson, An. Nat. Hist. 1859 (ser. 3, vol. 3) p. 269.—Pfeif. Malak. Blat. 1859, p. 23; Mon. Helic. vol. 5, p. 180.
Rungun valley near Darjiling.

3. **H. artificiosa**, Benson, An. Nat. Hist. ser. 2, vol. 18 (1856), p. 249.—Pfeif. Mon. Helic. vol. 4, p. 178 : Novit. vol. 1, pl. 36, f. 8, 9, 10.
Aracan.

1. 5. **H. climacterica**, Benson, J. Asi. Soc. Beng. 1856, p. 352 (amended in An. Nat. Hist. 1848, ser. 2, vol. 2), p. 163.—Reeve, Conch. Icon. Helix, f. 742.—Pfeif. Mon. Helic. vol. 3, p. 170; Kuster Chemn. Helix, pl. 141, f. 9, 10.
Terfa Ghat, Aracan Hills.
An imperforated species, with the whorls, which are dalt-ened above, somewhat scalariform.

6, 7. **H. Nilagirica**, Pfeiffer, Proc. Zool. Soc. 1845, p. 130 : Mon. Helic. vol. 1, p. 369.—Reeve, Conch. Icon. Helix, f. 450.
Nilgherries.

8, 9. **H. contracta**, Hutton MSS. in Benson An. Nat. Hist. ser. 3, vol. 12 (1864), p. 137.—Pfeif. Mon. Helic. vol. 5, p. 365.
Malwa.

10. **H. Indica**, Pfeiffer, Symb. Helic. pt. 3, p. 66 : Mon. Helic. vol. 1, p. 80.—Reeve, Conch. Icon. Helix, f. 448.
Nilgherries.

PLATE LVI.

HELIX.

1. **H. pansa**, Benson, An. Nat. Hist. ser. 2, vol. 18 (1856), p. 252.—Pfeif. Mon. Helic. vol. 4, p. 28.
Akoutong, Birmah.

2, 3. **H. catostoma**, Blanford, Proc. Zool. Soc. 1869, p. 447 (as section Trachia).
Ponsee in Yunan.

4 **H. subdecussata**, Pfeiffer, Proc. Zool. Soc. 1857, p. 107 : Mon. Helic. vol. 4, p. 28.
Bombay.

5, 6. **H. Koondaensis**, Blanford, J. Asi. Soc. Beng. 1870, vol. 39, pt. 2, p. 16 (and Cont. Ind. Mal. pt. 11), pl. 3, f. 12 (as Nanina).
Sispara in Koonda Hills.

7. **H. Baconi**, Benson, An. Nat. Hist. 1850 (ser. 2, vol. 6), p. 251.—Pfeif. Mon. Helic. vol. 3, p. 5 ; Kuster. Chemn. Helix, pl. 141, f. 11, 12.—Reeve, Conch. Icon. Helix, f. 1166.
Banks of Gungun, near Moradabad.
The unique type here delineated is decidedly immature, and closely approaches subdecussata and Layardi.

8, 9. **H. Layardi**, Pfeiffer, Proc. Zool. Soc. 1851 : Mon. Helic. vol. 3, p. 55.—Reeve, Conch. Icon. Helix, f. 614.
Ceylon.

10. **H. Emiliana**, Pfeiffer, Proc. Zool. Soc. 1852 ; Mon. Helic. vol. 3, p. 55; Kust. ed. Chemn. Helix, pl. 158, f. 33 to 38.—Not Reeve.
Ceylon.

———

PLATE LVII.

HELIX.

1, 2, 3. **H. anax**, Benson, An. Nat. Hist. 1865 (ser. 3, vol. 14).—Pfeif. Mon. Helic. vol. 5, p. 399.
Hills of Travancore.
The sculpture is close and obliquely concentric. There is a second more deeply-seated parietal lamella below the upcurving one, which the position selected by our artist did not enable him to exhibit.

4, 5, 6. **H. odontophora**, Benson, An. Nat. Hist. 1865 (ser. 3, vol. 14), Feb. p. 175.
From mountains 4500 feet high, Bandarewella and Bibiligamua, Ceylon.
Four remote palatal lamellæ can be descried from the exterior.

8, 9. **H. achatina**, Gray. See plate xiii. f. 1.

7, 10. **H. brachyplecta**, Benson, An. Nat. Hist. 1865, ser. 3, vol. 6, p. 319.
Banks of the River Attaran, near Moulmein.

PLATE LVIII.

HELIX.

1. **H. semidecussata**, Pfeiffer, Proc. Zool. Soc. 1851, p. 252: Kust. ed. Chemn. Helix, pl. 145, f. 8, 9.—Reeve, Conch. Icon. Helix, f. 567.
Ceylon.
Essentially identical with the Mauritian species.

2. **H. semidecussata**, Benson, var. solida.
Ceylon.
The solid and abnormal form here delineated has lost its epidermis, and almost unites semidecussata and Rosamonda.

3. **H. Laidlayana**, Benson, An. Nat. Hist. ser. 2, vol. 18 (1856), p. 253.—Pfeif. Mon. Helic. vol. 4, p. 51.—H. parietalis, Martens, Mal. Blät. 1864, p. 167, teste Pf.
Bengal.

4, 5. **H. Laidlayana**, Benson, var.
Cuttack (fig. 4).

6. **H. Peguensis**, Benson, An. Nat. Hist. 1860 (ser. 3, vol. 6), p. 192.—Pfeif. Mon. Helic. vol. 5, p. 346.
Sheeway Gheen, Pegu.

7, 8. **H. uter**, Theobald, J. Asi. Soc. Beng. 1859 (vol. 28, p. 309): Desc. Birm. p. 1 (separate pamphl.).—Pfeif. Mon. Helic. vol. 5, p. 127.
Near Moulmein.

PLATE LIX.

HELIX.

1, 2. **H. vitellina**, Pfeiffer, Proc. Zool. Soc. 1848, p. 109: Mon. Helic. vol. 3, p. 72; Kust. ed. Chemn. pl. 122, f. 22, 23.—Reeve, Conch. Icon. Helix, f. 390.
Nilgherries.

3. **H. Tranquebarica**, Fabricius, in Beck's Index (name only) and in Pfeif. Mon. Helic. vol. 1, p. 41.—Reeve, Conch. Icon, Helix, f. 394.
Tranquebar.

4. **H. semirugata**, Beck, Index Moll. Crist. Prod. p. 42 (name only), for H. globulus of Chemnitz, vol. 9, f. 1159, 1160, not of Müller.—Pfeif. Mon. Helic. vol. 1, p. 10; Kust. ed Chemn. Helix, pl. 3, f. 1112.—Reeve, Conch. Icon. Helix, p. 391.
Bengal.

5, 6. **H. Rosamonda**, Benson, An. Nat. Hist. 1860 (ser. 3, vol. 5), p. 381.—Pfeif. Mon. Helic. vol. 5, p. 77.
Pittewelle, Ceylon.

7, 8. **H. Theodori**, Philippi, Zeitsch. Malak. 1846, p. 191.—Pfeif. Mon. Helic. vol. 1, p. 70: Kuster ed. Chem. Helix, pl. 110, f. 1, 2, 3.—Reeve, Conch. Icon. Helix, f. 1188.
Birmah.

———

PLATE LX.

HELIX.

1, 2, 3. **H. Blanfordi**, Theobald, J. Asi. Soc. Beng. 1859 (vol. 28), p. 313: Desc. Birm. p. 4 (separate pamphlet).—Pfeif. Mon. Helic. vol. 5, p. 219.
Near Darjiling.
Very near H. cyclophax.

4. **H. ornatissima**, Benson, An. Nat. Hist. 1859 (ser 3, vol. 3), p. 269.—Pfeif. Mon. Helic. vol. 5, p. 113.
Pankalari, near Darjiling.

5. **H. capessens**, Benson, An. Nat. Hist. 1856 (ser. 2, vol. 18), p. 250.—Pfeif. Mon. Helic. vol. 4, p. 194: Novit. vol. 1, pl. 36, f. 17 to 20.
Moulmein, Birmah.

6. **H. bidenticulata**, Benson, An. Nat. Hist. 1852 (ser. 2, vol. 9), p. 465.—Reeve, Conch. Icon. Helix, f. 1184.—Pfeif. Mon. Helic. vol. 3, p. 165.
Nilgherries.

7. **H. crinigera**, Benson, An. Nat. Hist. 1850 (ser. 2, vol. 5), p. 214.—Reeve, Conch. Icon. Helix, f. 746.—Pfeif. Mon. Helic. vol. 3, p. 112; vol. 4, p. 110.

8. **H. diplodon**, Benson, An. Nat. Hist. 1859 (ser. 3, vol. 3), p. 187.—Pfeif. Malak. Blät. 1859, p. 13; Mon. Helic. vol. 5, p. 256.
Teria Ghat.

9, 10. We found this Helix in Benson's collection as H. Ingrami of Blanford, but know not whether it has been published.
Arracan Hills.

28 CONCHOLOGIA INDICA.

PLATE LXI.

HELIX.

1. **H. rimicola**, Benson, An. Nat. Hist., 1859 (s. 3, vol. 3), p. 266.—Pfeif. Mon. Helic., vol. 5, p. 71.
Near Lamlour, in W. Himalayah : a variety from Rungun, near Darjiling.

2, 3. **H. bullula**, Hutton, J. Asi. Soc. Beng., vol. 7, pt. 1 (1838), p. 2.—Pfeif. Mon. Helic., vol. 1, p. 86. —Not Reeve.
Lamlour ; Simla.

4, 5, 6. **H. humilis**, Hutton, J. Asi. Soc. Beng., vol. 7, pt. 1 (1838), p. 217.—Pfeif. Mon. Helic., vol. 1, p. 106; vol. 3, p. 83.—Reeve, Conch. Icon. Helix, f. 825.
Simla.

7, 8, 9. **H. nana**, Hutton, J. Asi. Soc. Beng., vol. 7, pt. 1 (1838), p. 218.—Pfeif. Mon. Helic., vol. 1, p. 31.
Simla.

10. **H. phyllophila**, Benson, An. Nat. Hist., 1863 (s. 3, vol. 11), p. 320.—Pfeif. Mon. Helic., vol. 5, p. 87.
Ceylon.

PLATE LXII.

HELIX.

1, 2, 3. **H. exul**, Theobald, J. Asi. Soc., Beng., 1864 (vol. 33), p. 245.
Andamans.

4, 5, 6. **H. stephus**, Benson, An. Nat. Hist., 1861 (ser. 3, vol. 7), p. 84.—Pfeif. Mon. Helic., vol. 5, p. 105.
Port Blair, Andamans.

7, 8, 9. **H. aspides**, Benson, An. Nat. Hist., 1863 (ser. 3, vol. 3), p. 320.—Pfeif. Mon. Helic., vol. 5, p. 197.
Birmah ; Andamans?

10. **H. radicicola**, Benson, An. Nat. Hist., 1859 (s. 2, vol. 2), p. 161.—Pfeif. Mon. Helic., vol. 3, p. 219; Kust. ed. Chemn. Helix, pl. 141, f. 13, 14.—Reeve, Conch. Icon. Helix, f. 753.
Himalaya.

PLATE LXIII.

HELIX.

1, 2, 3. **H. sequax**, Benson, An. Nat. Hist., 1859, April (ser. 3, vol. 3), p. 270.—Pfeif. Mon. Helic., vol. 5, p. 118.
Darjiling ; Rungun.

4, 5, 6. **H. vesicula**, Benson (as Nanina), J. Asi. Soc., Beng., 1858, vol. 7, p. 216.—Pfeif. Mon. Helic., vol. 1, p. 48; vol. 3, p. 47; Kust. ed. Chemn. Helix, pl. 129, f. 21, 22.—Not Reeve.
Sotee Durga, and Rajmahal.

7, 8, 9. **H. locythis**, Benson, An. Nat. Hist., 1860, p. 246.—Pfeif. Mon. Helic., vol. 3, p. 47.—Reeve, Conch. Icon. Helix, f. 1161.
Rajmahal Hills.

10. **H. glauca**, Benson, in Pfeif. Symb. Helic., pt. 3, p. 65 (as Nanina) ; Mon. Helic., vol. 1, p. 48.—Philip. Ab. N. Conch., vol. 3 Helix, pl. 10, f. 8.—Reeve, Conch. Icon. Helix, f. 771.
Almorah, Bengal.

PLATE LXIV.

HELIX : Section Zonites.

1, 2, 3. **H. subjecta**, Benson, An. Nat. Hist., 1852, p. 407.—Pfeif. Mon. Helic. vol. 3, p. 48.—Reeve, Conch. Icon. Helix, f. 1165.
Rajmahal Hills.

4, 5. **H. todarum**, W. and H. Blanford, J. Asi. Soc. Beng. 1861, p. 352, pl. 2, f. 1.
Near Pykara and Neddiwuttom, Nilgherries.

6, 7. **H. hypoleuca**, Blanford, J. Asi. Soc. Beng. 1865, p. 67, and Cont. Mal. Ind., pt. 5, p. 3 (as Nanina).—Pfeif. Mon. Helic., vol. 5, p. 104.
Akontong, Pegu.

8, 9, 10. **H. nebulosa**, Blanford, J. Asi. Soc. Beng., 1865, p. 66, and Cont. Mal. Ind., pt. 5, p. 2 (as Nanina).
Akontong, Pegu.

PLATE LXV.

VITRINA.

1, 4. **V. Bonsoni,** Pfeiffer, Proc. Zool. Soc. 1848, p. 107 : Mon. Helic. vol. 2, p. 497.—Reeve, Conch. Icon. Vit. f. 9.
Howrah, near Calcutta.

2, 3. **V. Peguensis,** Theobald, J. Asi. Soc. Beng. vol. 33 (1864), p. 244 : separate pamphlet, p. 8.
Near Pegu.

5, 6. **V. præstans,** Gould, Proc. Bost. Soc. N. H. vol. 1 (1843), p. 140 ; Boston J. Nat. H. vol. 4, p. 456, pl. 24, f. 2.—Pfeif. Mon. Helic. vol. 2, p. 497.—Reeve, Conch. Icon. Vit. f. 12.
Tavoy.
Our figure 6 is scarcely round enough.

7, 10. **V. succinea,** Reeve, C. Icon. Vitr. f. 8, for **V. planospira,** Benson (not Pfeiffer), An. Nat. Hist. ser. 3, vol. 3 (1859), p. 271, from which Pfeif. Mon. Helic. vol. 5, p. 14.
Punkabari ; Rangoon ; Khasia Hills.

8, 9. **V. Salius,** Benson, An. Nat. Hist. ser. 3, vol. 3 (1859), p. 189.—Pfeif. Mon. Helic. vol. 4, p. 792.
Khasia Hills.
Our figure, from the position selected by the artist, does not seem deep enough on the body-whorl.

PLATE LXVI.

VITRINA.

1, 4. **V. scutella,** Benson, Am. Nat. Hist. ser. 3, vol. 3 (1859, March).—Pfeif. Mon. Helic. vol. 4, p. 798.—Reeve, C. Icon. Vitr. f. 13 (from Benson's specimen).
Teria Ghat ; Khasia Hills ; Cashmire (var).

2, 3. **V. gigas,** Benson, J. Asi. Soc. Beng. vol. 5, (1836), p. 350.—Pfeif. Mon. Helic. vol. 2, p. 496.—Reeve, Conch. Icon. Vit. f. 3.
Sylhet ; Cherra, above Teria Ghat.

5, 6. **V. Flemingiana,** Pfeiffer, Proc. Zool. Soc. 1856, p. 324 : Mon. Helic. vol. 4, p. 790 ; Novit. vol. 1, pl. 28, f. 1 to 3.—Reeve, Conch. Icon. Vit. f. 4.
Scinde.

7, 10. **V. Christianæ,** Theobald, J. Asi. Soc. Beng. vol. 33 (1864), p. 245 : separate pamphlet, p. 9.
Andaman Isles.

8, 9. **V. irradians,** Pfeiffer, Proc. Zool. Soc. 1852, p. 156 : Mon. Helic. vol. 3, p. 3.—Reeve, Conch. Icon. Vit. f. 5.
Ceylon.

PLATE LXVII.

SUCCINEA.

1, 4. **S. Indica,** Pfeiffer, Proc. Zool. Soc. 1849, p. 133 : Mon. Helic. vol. 3, p. 8.
Bheemtal.

2, 3. **S. semiserica,** Gould, Proc. Bost. Soc. Nat. H. (1816), vol. 2, p. 100.—Pfeif. Mon. Helic. vol. 3, p. 10.
Tavoy, Birmah.

5, 6. **S. Girnarica,** Theobald, J. Asi. Soc. Beng. 1859, vol. 28, p. 309 : Desc. Burm. Hd. p. 5.—Pfeif. Mon. Helic. vol. 4, p. 792.
Girnar Hills, Gujerat.

7. **S. daucina,** Pfeif. Proc. Zool. Soc. 1854, p. 298 : Mon. Helic. vol. 4, p. 810.
Calcutta.

8. **S. plicata,** Blanford, J. Asi. Soc. Beng. 1865, pt. 2, (vol. 34), p. 80 : Cont. Mal. Ind. pt. 5.
Tongoop, Aracan ; Pegu (var.).

9. **S. Bensoni,** Pfeiffer, Proc. Zool. Soc. 1849, p. 133 : Mon. Helic. vol. 3, p. 9.
Moradabad.

10. **S. rutilans,** Blanford, J. Asi. Soc. Beng. 1870, vol. 39, pt. 2, p. 23, pl. 3, f. 23.
Cherra Punji.

PLATE LXVIII.

SUCCINEA.

1, 4. **S. Baconi,** Pfeiffer, Proc. Zool. Soc. 1851, p. 298 : Mon. Helic. vol. 4, p. 804.
Calcutta.

2, 3. **S. vitrea,** Pfeiffer, Proc. Zool. Soc. 1854, p. 298 : Mon. Helic. vol. 4, p. 810.
Calcutta.

5, 6. **S. crassiuscula**, Benson's Mss. in Pfeif. Mon.
Helic. vol. 3, p. 9.
Bundelkhund : Punjaub.

7. **S. acuminata**, Blanford, Proc. Zool. Soc. 1869,
p. 449.
Moenein in Yunan.

8, 9. **S. collina**, Blanford, Mss.
Mahabaleshwar.

10. **S. collina**, var.

PLATE LXIX.

LIMNÆA.

1, 4. **L. rufescens**, Gray, in Sowerby's Genera
Shells, pt. 7, Limn. f. 2, and Reeve, Conch. System.
pl. 191, f. 2.—L. chlamys, Benson, in part.
Ganges, &c., &c.

The name is infelicitous, because the reddish tint is
rather abnormal than otherwise : it has however a long
priority of date. Hereafter links may be discovered
to unite the species with the still earlier-named acuminata. We suspect that eventually all the Indian
forms (those of the Germanic region excepted) will be
referred to acuminata, luteola, and ovalis.

2, 3. **L. rufescens**, var. patula.
Ganges.

Apparently the L. patula of Troschel in Wiegmann's
Archives for 1837 (vol. 3, p. 167).

5, 6. **L. chlamys**, Benson, Journ. Asi. Soc. Beng.
1836, vol. 5, p. 744.
Moradabad, Benares, &c.

Runs into rufescens, which was apparently co-
extensive as a Bensonian species. Yet as the links
have not been obtained by us, this extreme form
(from Benson's collection) may for convenience sake
retain his appellation.

7, 10. **L. amygdalus**, Troschel, Wiegm. Archiv. 1837,
vol. 3, p. 168.—Kuster, ed. Chemn. Lim. p. 35,
pl. 6, f. 15, 16.
Ganges.

Perhaps only a variety of rufescens, but the sutural
line is not so oblique, and the colour differs.

8, 9. **L. acuminata**, Lamarck, Anim. s. Vert. vol. 6,
pt. 2, p. 160.—Delss. Rec. Coq. Lam. pl. 30, f. 6.
30 miles S.E. of Hingola, &c.
Lamarck's Bengal specimens were probably rufescens,

but the shape of the individual delineated from his
cabinet accords better with that which we have
figured.

PLATE LXX.

LIMNÆA.

1. **L. rufescens**, Gray, var.
Some regard this abnormal form as the L. pectinoides
of Kuster's monograph.

2, 3. **L. ovalis**, Gray, in Sow. Gen. Shells, pt. 7,
Limn. f. 4, Reeve, Conch. Syst. pl. 191, f. 4.
Calcutta ; Almorah.

4. **L. ovalis**, Gray, var. strigata.
Jounpore.
Has the aspect of L. ceraserum of Troschel.

5, 6. **L. luteola**, Lamarck, Anim. s. Vert. vol. 6, pt. 2,
p. 160.—Delss. Rec. Coq. Lam. pl. 30, f. 5.
Bengal, &c., &c.
This and rufescens seem diffused throughout India

7, 10. **L. pinguis**, Dohrn, Proc. Zool. Soc. 1858,
p. 134.
Ceylon.

8. **L. pinguis**, Dohrn, var.
Calcutta.

Benson's types of his L. bulla (misprinted butta), a
mere name for the almost undescribed L. limosa ? of
Hutton (J. Asi. Soc. Beng. vol. 3, 1834) chiefly belong
to this form. The L. bulla of Kuster (ed. Chemn.) is
more like the European ovata (peregra var. ovata).
Our specimen almost unites pinguis with luteola.

9. **L. rufescens**, Gray, var. Sylhetica.
Marshes in Sylhet.

A rare form delineated from Benson's original type
of the Sylhet variety of his chlamys.

PLATE LXXI.

MELANIA.

1. **M. Iravadica**, Blanford, Proc. Zool. Soc. 1869,
p. 445.
Upper Irawady at Malé and Bhamo.

2, 3. **M. Broti**, Dohrn, in Reeve's Conch. Icon.
Mel. f. 160.—M. chocolatum, Brot, Revue Zool.
1860, June, pl. 16, f. 2.
Ceylon.

4. **M. zonata**, Benson, Journ. Asi. Soc. Beng. 1836, vol. 5, p. 747.—Philippi, Ab. N. Conch. vol. 1, Mel. pl. 1, f. 12.—Reeve, Conch. Icon. Mel. f. 217.
Sylhet.

5, 6. **M. Hugeli**, Philippi, Ab. N. Conch. vol. 1, p. 61, Mel. pl. 2, f. 8.—M. siphonata, Reeve, Conch. Icon. Melan. f. 143.
Khasia Hills ; Mysore.

7. **M. lineata**, Gray, Index Test. Sup. (1828), Helix, f. 68.—Trosch. in Wiegm. Arch. Nat. 1837, p. 176.—M. lirata, Benson, Journ. Asi. Soc. Beng. 1836, vol. 5, p. 782, name only for nameless fig. D in Gloan. Sci. Calcut. vol. 1 (1829).—Reeve, Conch. Icon. Mel. f. 170.
River Goomty ; Tenasserim, &c.

8, 9. **M. terebra**, Benson, Journ. Asi. Soc. Beng. 1836, vol. 5, p. 747. — Reeve, Conch. Icon. Mel. f. 9.—M. torquata, Busch. in Philippi Ab. N. Conch. vol. 1, Mel. pl. 1, f. 18.
Sylhet.

10. **M. Riqueti**, Grateloup, Trans. Lin. Bordeaux, vol. 11, pl. 5, f. 28.
Quilon, Travancore ; Cochin.

According to Brot, the tornatella of Reeve, supposed by him to be identical, is not this species : his *b* is the veritable tornatella of Lea ; his *a* seems the sculpta of Souleyet (Zool. Bonite), a near ally.

PLATE LXXII.

MELANIA.

1, 2. **M. gloriosa**, Anthony, Americ. J. of Conch. vol. 1 (1865), pt. 3, p. 207, pl. 18, f. 2.
Bassein district, Pegu.

3. We had proposed the name of Goliah for this magnificent species, but as it is evidently identical with some smaller specimens which may possibly (?) prove the humerosa of Gould, or the infrapicta of Martens (so termed in Cuming's collection), we defer the naming of it until the next part of our publication.

4. **M. confusa**, Dohrn, Proc. Zool. Soc. 1858.
Ceylon.

Closely allied to the Tirouri of Quoy, and the erosa of (" Lesson ") Philippi.

5. **M. Herculea**, Gould, Proc. Bost. Soc. Nat. II. 1846, vol. 2, p. 110 : Otia Conch. p. 199.
Tavoy River.

Reeve (Conch. Icon. Mel. f. 4) has figured an Herculea from Ceylon ! which is utterly unknown to us from that quarter.

6. **M. Peguensis**, Anthony, Amer. Journ. Conch. Pegu.

7. **M. episcopalis**, Lea, Proc. Zool. Soc. 1850, p. 184, in part.
Diyung River, North Cachar Hills.

See plate 75, f. 5, 7.

PLATE LXXIII.

MELANIA.

1, 2, 3, 4. **M. scabra**, Müller (as Buccinum s.), Hist. Verm. vol. 2, p. 136. — Chemn. Conch. Cab. f. 1259, 1260, badly, (as Helix s.).—Brug. Enc. Méth. Vers, vol. 1, p. 330 (as Bulimus s.).—Desh. ed. Lam. Anim. s. Vert. vol. 9, p. 443.— Helix aspera, Gmel. Syst. Nat. 3656.—Dillw. Cat. vol. 2, p. 950.—Wood, Ind. Testac. pl. 34, f. 141.—M. elegans, Reeve, Conch. Icon. Mel. f. 178.
Poona ; Coromandel ; Cochin ; River Goomti ; Ceylon (fig. 3, 4), &c., &c.

A very variable shell, of which the coarser and more spinous form (f. 1) comes from Cochin and Beloochistan ; the smoother and less opaque from Poona.

5, 6, 7. **M. scabra**, var. elegans.—M. elegans, Bens. Journ. Asi. Soc. Beng. vol. 5 (1836), p. 782, name only for the unnamed turreted form in Gleanings in Science, vol. 1 (1829), Melan. letter c.—Hutton, Journ. Asi. Soc. Beng. (vol. 17, pt. 1) 1849, p. 657.
North of Oude, S. India.

Figure 6 is drawn from a shell marked by Benson as his type ; figure 5 is peculiarly characteristic ; figure 7 is an elongated variety, from the North of Oude.

8, 9. **M. Layardi**, Dohrn, Proc. Zool. Soc. 1858.— Reeve, Conch. Icon. Mel. f. 104.
Ceylon.

10. **M. datura**, Dohrn, Proc. Zool. Soc. 1858.— Reeve, Conch. Icon. Mel. f. 213.
Ceylon.

PLATE LXXIV.

MELANIA.

1, 2, 3, 4. **M. tuberculata**, Müller, Hist. Verm. pt. 2, p. 191 (as Nerita t.) and in Chenn. Conch. Cab. vol. 9, f. 1261, 1262.—Philippi, Ab. N. Conch. vol. 1, Mel. pl. 1, f. 19.—Bulimus t. Bruguière, Enc. Méth. Vers, vol. 1, p. 350.—Stroubus vibex, Gmel. Syst. Nat. p. 3522.—Dillwyn, Desc. Cat. Sh. vol. 2, p. 950. —Melanoides fasciolata, Olivier, Voy. pl. 31, f. 7.
Abundantly diffused : the finest from Southern India.

5, 6. **M. Tirouri**, var. ? Férussac in Quoy and Gaim. Voy. Astrol. Zool. vol. 3, p. 159, pl. 56, f. 38, 39.
Paniar (or Pannaar) River, Cuddalore, S. India.
The only adult example known to us is the one here figured. It may possibly prove distinct from the species to which we doubtfully refer it, but we dare not delineate it as new. Perhaps it may be the Helix turrita of Chemnitz (Conch. Cab. vol. 9, pt. 2, p. 6, for H. turrita crenulata, p. 165, f. 1256), but it is devoid of the infrasutural crense.

7, 10. **M. rudis**, Lea, Proc. Zool. Soc. 1850, p. 186. —Reeve, Conch. Icon. Mel. f. 172.
Ceylon.
A very close approach to the spineless form of Broti. The M. microstoma of Lea, ascribed to Ceylon by Brot (not by Lea), and considered identical by some writers, does not agree with the specimen here delineated.

8, 9. **M. Batana**, Gould, Proc. Bost. Soc. Nat. H. vol. 1, p. 141 : Otia Conch. p. 191.
Tenasserim.

PLATE LXXV.

MELANIA.

1. 4. **M. baccata**, Gould, Proc. Bost. Soc. Nat. Hist. vol. 2, p. 219; Otia Conch. p. 200.
Thoungyin River, Birmah.

2. **M. baccata**, Gould, var. fusiformis.
Shan States.

3. **M. baccata**, Gould, var. pyramidalis.—M. variabilis, var. pyramidalis, Theobald, Journ. Asi. Soc. Beng. vol. 34, pt. 2 (1865), pl. 19, f. 7.
Shan States.

5. 7. **M. episcopalis**, Lea, Proc. Zool. Soc. 1850, p. 181, in part.—Reeve, Conch. Icon. Mel. f. 12.
Diyung River, North Cachar.
This is not the episcopalis of the Conchological Mis-

cellany (from Borneo), which, formerly confused with it, has been been termed Brookei by Reeve (C. Icon Mel. f. 207).

6. **M. variabilis**, var. spinosa.—M. spinosa, Benson, in Hanley's Conch. Miscel. Mel. pl. 1, f. 7 (small form).—M. variabilis, var. B. Benson, Journ. Asi. Soc. Beng. vol. 5, p. 746.
River Jumna; Syllet, &c.
The spire in perfect examples of this rare form is wont to be quite smooth : the character, however, is not invariable.

PLATE LXXVI.

PALUDINA.

1, 4. **P. Naticoides**, Theobald, Journ. Asi. Soc. Beng. vol. 34, 1865, pt. 2, pl. 9, f. 1, 2, 3.
Shan States.

2, 3. **P. crassa**, Hutton Mss. in Benson Journ. Asi. Soc. Beng. vol. 5, 1836, p. 745.—Reeve, Conch. Icon. Palud. f. 33.—P. obtusa, Troschel, Wiegm. Arch. Nat. Hist. 1837, p. 173.—Philippi, N. Conch. vol. 1, p. 116, Pal. pl. 1, f. 14.—Reeve, Conch. Icon. Palud. f. 32.
Bengal.
Varies much in elevation of spire.

5. **P. oxytropis**, Benson, Jour. Asi. Soc. Beng. vol. 5 (1836), p. 745.—Reeve, Conch. Icon. Palud. f. 9.— P. pyramidata, Philippi, Ab. N. Conch. vol. 1, Pal. pl. 1, f. 3, 4.—Kust. ed. Chemn. Palud. pl. 6, f. 1, 2.
Bengal (teste Reeve and Philippi).
Still very rare : figured from the original type.

6. **P. lecythis**, Benson, Journ. Asi. Soc. Beng. vol. 5 (1836), p. 745.
Upper Birmah.
We have figured the almost unique type.

7. **P. lecythis**, var. ampulliformis, Benson.—P. ampulliformis, Eydoux and Souleyet, Voy. Bonite, Zool. p. 549, pl. 31, f. 25, 26, 27.
Upper Birmah.
Only differs from the type by the absence of the infrasutural angle.

8, 9, 10. **P. Bengalensis**, Lamarck, Anim. s. Vert. (ed. Desh.) vol. 8, p. 513.—Deles. Rec. Coq. Lam. pl. 31, f. 2.—Reeve, Conch. Icon. Palud. f. 5.—Kust. ed. Chemn. Palud. f. 15, 16.—P. elongata, Swains. Zool. Ill. ser. 1, pl. 98, top.—P. lineata, Valenc. in Humb. & Bonpl. Voy. Zool. vol. 2, p. 255.
Ganges, &c.

Under this name Deshayes has figured a shell in Bélanger's voyage (Ind. Orient. Zool. p. 419, Moll. pl. 1, f. 14, 15) which differs remarkably from the ordinary type. Our figure 10 is from a rather uncommon variety.

PLATE LXXVII.
PALUDINA.

1, 2. **P. Ceylanica**, Dohrn, Proc. Zool. Soc. 1857, p. 123.—Reeve, Conch. Icon. Palud. f. 32.
Ceylon.
Possibly identical with the P. biangulata of Kuster.

3, 4. **P. dissimilis**, Müller, Hist. Verm. pt. 2, p. 184 (as Nerita?.)—Schröter, Einleit. Conch. vol. 2, p. 254, pl. 4, f. 10, (ditto).—Helix d. Gmelin, Syst. Nat. 3647.—Dillwyn, Des. Cat. p. 941 (ditto).— P. Remossii, Kuster, (not Philippi) ed. Chemn. Palud. p. 26, pl. 5, f. 17, 18.
Tanks near Calcutta ; Koodooruwave, &c.
This shell was fairly figured by Chemnitz as the Tranquebar variety of the Helix vivipara of Linnæus. The P. dissimilis of Reeve was not Indian, but from the Nile, and is not unlike unicolor : his melanostoma (C. Icon. Pal. f. 27) is regarded by Martens as a form of this species.

5. **P. Bengalensis**, Lamarck, var. gigantea.—P. gigantea, Von dem Busch, in Reeve's Conch. Icon. Palud. f. 7.
Bengal.
A swollen form of this widely diffused species.

6. **P. doliaris**, Gould, Proc. Bost. Soc. Nat. Hist. vol. 1, p. 144 ; Otia Conch. p. 191.—Reeve, Conch. Icon. Palud. f. 1.
British Birmah.

7, 10. **P. Heliciformis**, Frauenfeld, Verhandl. Zool. Bot. Wien, vol. 15 (1865), p. 533, pl. 22 (as Vivivipara H.) : Zool. Misc. pt. 5.—P. dissimilis, var. decussatula, or P. decussatula, Blanford, Proc. Zool. Soc. 1869, p. 446.
Ava ; Rangoon.
Comparison of the types shows their absolute identity. The specimen originally described was supposed to have come from Central Africa. The spire is usually eroded.

8, 9. **P. Remossii**, Philippi (erroneously as of Benson), Abbild. N. Conch. vol. 2, p. 134, Palud. pl. 2, f. 3.
Joanpore, Soobathur, &c.
We are aware that Benson supposed "Remossii" was an incorrect reading of his manuscript name

"primorsa," yet as the species actually published under the latter designation by Reeve is much more like dissimilis, and the shell here delineated was the melanostoma of Hutton according to Benson's collection, we prefer to retain the printed name. It matters little, indeed, to science whether or not a Monsieur Remossæ ever existed ; much to science that published names should not be changed without the most absolute necessity. The P. melanostoma of Reeve (?— dissimilis) does not exhibit the dark-lipped mouth ; no definition of Hutton's shell has, to our knowledge, appeared in print.

PLATE LXXVIII.
ACHATINA : Section Electra.

See previous plates, xvii, xviii, xxxv, xxxvi.

1. **A. pulla**, Blanford, Journ. Asi. Soc. Beng. 1870, vol. 39, pt. 2, p. 21, pl. 3, f. 20 (as Glessula p.)
Toma.

2. **A. Hugeli**, Pfeiffer, Mon. Helic. vol. 2, p. 259.— Glandina H. Philip. Ab. Neuer Conch. vol. 1, p. 135, Glandina, pl. 1, f. 8.
Cashmire.

3. **A. Tornensis**, Blanford, Journ. Asi. Soc. Beng. 1870, vol. 39, pt. 2, p. 22, pl. 3, f. 22 (as Glessula.)
Torna Hill, near Poona, Deccan.

4. **A. Sattaraensis**, H. Adams, Mss. for his A. fusca, Proc. Zool. Soc. 1868, p. 15, pl. 4, f. 10 (preoccupied).
Saharunpore, Ceylon.
This is not the A. fusca of Pfeiffer, which is near, if not identical with the paraldilis of Benson.

5. **A. crosa**, Blanford, Journ. Asi. Soc. Beng. 1871, (vol. 40, pt. 2), p. 43, pl. 2, fig. 7 (as Glessula c.).
Darjiling.

6. **A. baculina**, Blanford, Journ. Asi. Soc. Beng. 1871, vol. 40, pt. 2, p. 43, pl. 2, f. 6 (as Glessula b.).
Khersiong, Sikkim Himalayah.

7. **A. Singhurensis**, Blanford, Journ. Asi. Soc. Beng. 1870, vol. 39, pt. 2, p. 19, pl. 3, f. 17 (as Glessula S.).
Singhur, near Poona, Deccan.

8. **A. serena**, Benson, Ann. Nat. Hist. 1860, ser. 3, vol. 5, p. 394, 460.—Pfeif. Mon. Helic. vol. 6, p. 223.
Ceylon.

9. **A. Oreas**, Benson, in Reeve's Conch. Icon. Achat. f. 113.—Pfeif. Mon. Helic. vol. 3, p. 495 ?
Nilgherries.

The single and slightly broken shell figured by Reeve must be regarded as the type, but as Pfeiffer's description does not well apply to it, and Benson had another species mixed with it, it is probable the specimen he lent Pfeiffer was not the original.

10. **A. Jerdoni,** Benson, in Reeve's Conch. Icon. Ach. f. 80.—Pfeif. Mon. Helic. vol. 3, p. 495. Nilgherries.

PLATE LXXIX.

SPIRAXIS, &c.

See previous plates xix to xxiii.

1. **S. Cingalensis,** Benson, Ann. Nat. Hist. 1863, Feb. (ser 3, vol. 11) p. 91.—Pfeif. Mon. Helic. vol. 6, p. 191.

Woolgunowe, Mattelle, Ceylon.

The apex of the unique example (which looks like a Syrnola) is broken off. Under Spiraxis authors have grouped most utterly different forms.

2, 3. **S. Layardi,** Benson, Ann. Nat. Hist. (ser. 3, vol. 11) 1863, Feb. p. 90.—Pfeif. Mon. Helic. vol. 6, p. 190.

Ceylon.

4. **S. Walkeri,** Benson, Ann. Nat. Hist. 1863, (ser. 3, vol. 11) p. 90.—Pfeif. Mon. Helic. vol. 6, p. 189.

Port Blair, Andamans.

5. **S. Haughtoni,** var. Benson.

The Opeas Pealei of Tryon (Americ. J. of Conch. vol. 5, p. 110, pl. 10, f. 5) is an additional synonym of this rare species.

6. **Bulimus densus,** Pfeiffer, Malak. Blat. 1855, p. 144: Mon. Helic. vol. 4, p. 424.

Malabar.

7. **B. latebricola,** Benson, in Reeve's Conch. Icon. Bulim. f. 572.—Pfeif. Mon. Helic. vol. 3, p. 401: Kust. ed. Chemn. Bul. pl. 20, f. 5, 6.

Landour, W. Himalayah.

8. **S. pusillus,** Blanford, Journ. Asi. Soc. Beng. vol. 34, 1865, p. 78 : Cont. Mal. pt. 5.—Pfeif. Mon. Helic. vol. 6, p. 192.

Prome, Pegu.

9. **B. scrobiculatus,** Blanford, Journ. Asi. Soc. Beng. 1865 (vol. 34), p. 77: Cont. Mal. pt. 5, p. 13. —Pfeif. Mon. Helic. vol. 6, p. 151.

Pegu, west of the Irawady.

10. **S. hobos,** W. and H. Blanford, Journ. Asi. Soc. Beng. 1861, p. 361, pl. 1, f. 15 : Cont. Mal. pt. 2. —Pfeif. Mon. Helic. vol. 6, p. 190.

Nilgherries.

PLATE LXXX.

BULIMUS.

See previous plates xix to xxiii.

1. **B. cœlebs,** Benson's Ms. in Pfeif. Symbol. pt. 3, p. 83 : Mon. Helic. vol. 2, p. 119. — Reeve, Conch. Icon. Bulim. f. 301.

Near Almorah, Bengal; Landour and Kemaon, W. Himalayah.

2. **B. cœlebs,** var. ceratina.—B. ceratinus, Bens. in Reeve's Conch. Bulim. f. 569.

Almorah ; Kemaon, W. Himalayah.

3. **B. prolotarius,** Pfeiffer, Proc. Zool. Soc. 1854, p. 292 : Mon. Helic. vol. 4, p. 417.

Ceylon.

4. **B. trutta,** Blanford, Journ. Asi. Soc. Beng. vol.36, 1866, p. 42 : Cont. Mal. pt. 6, p. 12.— Pfeif. Mon. Helic. vol. 6, p. 125.

Anamullay Hills.

5. **B. Bontiæ,** Chemnitz, Conch. Cab. vol. 9. f. 1216, 1217 (as Helix).—Pfeif. Mon. Helic. vol. 2, p. 194 : Kust. ed. Chemn. Bul. pl. 10, f. 8, 9.

Southern India.

6. **B. lepidus,** Gould, Proc. Bost. Soc. Nat. H. 1856, p. 11 : Otia Conch. p. 219.—Pfeif. Mon. Helic. vol. 6, p. 33.

Mergui Isles.

Except in being shorter, with more rounded whorls, and without a notch at the commencement of the pillar-lip, it might be taken for Syphoticus.

7. **B. Bengalensis,** Lamarck, Anim. s. Vert. (ed. Desh.) vol. 8, p. 233.—Deles. Rec. Coq. Lam. pl. 28, f. 1.—Pfeif. Mon. Helic. vol. 2, p. 194.— Reeve, Conch. Icon. Bulim. f. 289.

Bengal.

8. **B. plicifer,** Blanford, Journ. Asi. Soc. Beng. vol. 34, 1865, pt. 2, p. 78 : Cont. Mal. pt. 5.— Pfeif. Mon. Helic. vol. 6, p. 102.

Prome, Pegu.

9. **B. putus,** Benson, Ann. Nat. Hist. ser. 2, vol. 19, 1857, April.—Pfeif. Mon. Helic. vol. 4, p. 502.

Tavoy.

10. **B. segregatus,** Benson, in Reeve's Conch. Icon. Bulim. f. 587.

Simla, W. Himalayah ; Cashmire.

PLATE LXXXI.

ANCYLUS, CAMPTONYX, LITHOTIS, LITHOGLYPHUS.

1, 4. **A. Ceylanicus**, Benson, An. Nat. Hist. 1864, ser. 3, vol. 13, p. 139.
Matelle, Ceylon.

2, 3. **A. verruca**, Benson, An. Nat. Hist. 1855, ser. 2, vol. 15, p. 12.
Bhianthal, Rohilkhund, Orissa : Juwai, Cachar.

5, 6. **C. Theobaldi**, Benson, An. Nat. Hist. 1858, ser. 3, vol. 1, p. 396, pl. 12, f. 1, 2.—Valenciennesia T. Fischer, Jour. Conch. 1859, vol. 7, p. 319.
Mount Girnar, Kattiawar, W. India.

7. **Lithotis rupicola**, Blanford, An. Nat. Hist. 1863, ser. 3, vol. 12, p. 186, pl. 4, f. 8, 9, 10.
Bori Ghah.

8, 9. **L. tumida**, Blanford, Jour. Asi. Soc. Beng. 1870, vol. 39, pt. 2, p. 25, pl. 3, f. 24.
Singhur and Poorundjhur (subcostulate var.).

10. **Lithoglyphus Martabanensis**, Theobald, Jour. Asi. Soc. Beng. 1870, vol. 39, pt. 2, p. 402, pl. 18, f. 9.
Martaban.

PLATE LXXXII.

CYATHOPOMA.

1, 4. **C. Malabaricum**, W. and H. Blanford, J. Asi. Soc. Beng. 1860 (Cont. Mal. pt. 1, p. 9), as Cyclotus.—Pfeif. Mon. Pneum. vol. 3, p. 32 (as Cyclotus).—W. Blanford, J. Conch. 1868, vol. 16, p. 261, pl. 12, f. 7 (as Cyath.).
Near Pykara.
The colour is solely in the epidermis.

2, 3. **C. filocinctum**, Benson, An. Nat. Hist. 1851, ser. 2, vol. 8, p. 188 (as Cyclostoma).—Pfeif. Mon. Helic. vol. 1, p. 221 (as ? Cyclostomus) ; vol. 2, p. 25 (as Cyclotus).—Blanf. J. Conch. 1868, vol. 16, p. 258, pl. 12, f. 1.—Reeve, Conch. Icon. Cyclot. f. 50 (as Cyclotus).
Nilgherries, Southern India.

5, 6. **? C. malleatum**, Blanford, J. Asi. Soc. Beng. 1861 (vol. 30), p. 349, pl. 1, f. 6 (as Cyclophorus).—Pfeif. Mon. Pneumon. vol. 3, p. 71 (as Cyclop.).
Sherroy Hills, S. India.

We have doubtfully referred this shell to Cyathopoma, from its greater resemblance to the species here figured than to the Cyclophori : its operculum is unknown to us.

7, 10. **C. tignarium**, Benson, An. Nat. Hist. 1862, Dec. (ser. 3, vol. 12) p. 426 (as ? Cyath.).—Pfeif. Mon. Pneum. vol. 3, p. 33 (as Cyclotus).—Blanford, J. Conch. 1868, vol. 16, p. 263, pl. 12, f. 9.
Andaman Islands.

8, 9. **C. Deccanense**, Blanford, Journ. Conch. 1868, vol. 16, p. 258, pl. 12, f. 2.
Western Ghats, not far from Bombay.

PLATE LXXXIII.

HELIX.

See previous plates, xiii. to xvi. ; xxv. to xxxii. ; l. to lxiv.

1, 2, 3. **H. castra**, Benson, An. Nat. Hist. 1852 (ser. 2, vol. 10), p. 349.—Reeve, Conch. Icon. f. 1160.—Pfeif. Mon. Helic. vol. 3, p. 635.
Darjiling, Sikkim Himalayah.
The minute spiral striæ on the lower disk, near the keel, cannot be represented by lithography.

4, 7. **H. sanis**, Benson, An. Nat. Hist. 1861 (ser. 3, vol. 7), p. 84.—Pfeif. Mon. Helic. vol. 5, p. 186.
Port Blair, Andamans : Shan Provinces.
The lower disk has no spiral sculpture, but when very highly magnified seems rugosely shagreened. The keel seems too broad in figure 4.

5, 6. **H. galerus**, Benson, An. Nat. Hist. 1856 (ser. 2, vol. 18), p. 96.—Pfeif. Mon. Helic. vol. 4, p. 111.
Ceylon.
The base resembles that of H. castra.

8, 9, 10. **H. macromphalos**, Blanford, Jour. Asi. Soc. Beng. 1870, vol. 39, pt. 2, p. 17, pl. 3, f. 11 (as Plectopylis).
Marung in Khasi Hills : small var. from Darjiling.

PLATE LXXXIV.

HELIX.

1, 4. **H. pinacis,** Benson. See previous reference, pl. 13, f. 5.

2, 3. **H. Atkinsoni,** Theobald. See previous reference, pl. 15, f. 9.
The surface is adorned, both above and below, with very fine and close riblets, which, under a powerful glass, appear subarticulated.

5, 6. **H. Ataranensis,** Theobald, J. Asi. Soc. Beng. 1870 (vol. 39, pl. 2), p. 401, pl. 18, f. 7 (as Nanina).
Near the river Ataran, in Martaban.

7. **H. Gardeneri,** Pfeiffer, in Kust. ed. Chemn. Helix, pl. 112, f. 12, 13: Mon. Helic. vol. 1, p. 47.—Reeve, Conch. Icon. Helix, f. 416.
Ceylon.

8, 9, 10. **H. arata,** Blanford, Proc. Zool. Soc. 1869, p. 445 (as Nanina, section Rotula).
Near Bhamo, Birmah.

PLATE LXXXV.

HELIX.

1, 4. **H. convexa,** Reeve (as of Benson), Conch. Icon. Hel. f. 762, for H. monticola of Pfeif. (erroneously Hutton) in Kust. ed. Chemn. Hel. pl. 160, f. 3–5.
Himalayah.

2, 3. **H. ruginosa,** Férussac, Hist. Moll. pl. 71, f. 4 (Helicella).—Pfeif. Mon. Helic. vol. 1, p. 368: Kust. ed. Chemn. Helix, pl. 76, f. 7 to 10.—Reeve, Conch. Icon. Helix, f. 748.
Bengal.

5, 6. **H. partita,** Pfeif. Proc. Zool. Soc. 1853, p. 125: Mon. Helic. vol. 4, p. 55.—Reeve, Conch. Icon. Helix, f. 1311 (badly).
Ceylon.
Reeve's H. subopaca seems a mere variety.

7, 10. **H. subconoidea,** Pfeiffer, Proc. Zool. Soc. 1851, p. 51: Mon. Helic. vol. 4, p. 56, and vol. 5, p. 113.—Reeve, Conch. Icon. Helix, f. 1326.
Ceylon.

8, 9. **H. fallaciosa,** Férussac, Hist. Moll. pl. 71, f. 1, 2 (as Helicella).—Pfeif. Mon. Helic. vol. 1, p. 368.—Reeve, Conch. Icon. Helix, f. 459.
Coimbatore, Khoondah Hills: Ceylon.
Very like asperella, but is smooth.

PLATE LXXXVI.

HELIX.

1. **H. concavospira,** Pfeif. Proc. Zool. Soc. 1853, p. 124: Mon. Helic. vol. 4, p. 32.—Reeve, Conch. Icon. Helix, f. 1315.
Ceylon.

2, 3. **H. Zoroaster,** var. concolor, Theobald, Journ. Asi. Soc. Beng. 1859 (vol. 28), p. 310: Desc. Birm. p. 2.
Near the Irawadi, between Prome and Ava.

4. **H. turritella,** Adams (as Nanina), Proc. Zool. Soc. 1869, p. 275, for his Nanina conulus (preoccupied), Proc. Zool. 1867, p. 307, pl. 19, f. 16.
Ceylon.

5, 6. **H. Angelica,** Pfeif. Proc. Zool. Soc. 1856, p. 33: Mon. Helic. vol. 4, p. 123: Novit. pl. 21, f. 4, 5, 6.
Punjaub.

7. **H. Attegia,** Benson, An. Nat. Hist. 1859 (ser. 3, vol. 3), p. 184.—Pfeif. Mon. Helic. vol. 5, p. 91: Novit. pl. 78, f. 17, 18, 19.
Phie Than, Tenasserim.

8, 9, 10. **H. fritillata,** Benson, An. Nat. Hist. 1863 (ser. 3, vol. 11), p. 320.—Pfeif. Mon. Helic. vol. 5, p. 217.
Pegu.

PLATE LXXXVII.

HELIX.

1, 4. **H. orcula,** Benson, An. Nat. Hist. 1850 (ser. 2, vol. 6), p. 251.—Pfeif. Mon. Helic. vol. 3, p. 42. —Reeve, Conch. Icon. Hel. f. 1176.
Bengal, Behar, &c.

2, 3. **H. monoroma**, Benson, Au. Nat. Hist. 1853
(ser. 2, vol. 12), p. 92.—Pfeif. Mon. Helic. vol. 4,
p. 37.—Reeve, Conch. Icon. f. 1352.
Ceylon.

5, 6. **N. Pirriana**, Pfeif. Proc. Zool. Soc. 1854, p.
55 : Mon. Helic. vol. 4, p. 154, and vol. 6, p. 417.
—Reeve, Conch. Icon. Helix, f. 1541.
Walaghat in Koondah Hills.

7. **H. Barrakporensis**, Pfeiffer, Proc. Zool. Soc.
1852, p. 156 : Mon. Helic. vol. 3, p. 59, and vol. 5,
p. 86.
Barrakpore.

8, 9. **N. rotifera**, Pfeiffer, Proc. Zool. Soc. 1845, p.
13 : Mon. Helic. vol. 1, p. 119.—Reeve, Conch.
Icon. Helix f. 1170.—Blanford, An. Nat. Hist.
1864 (ser. 3, vol. 7), p. 244 (as Plectopylis).
Nilgherries.

10. **N. macroplouris**, Benson, Au. Nat. Hist. 1859
(ser. 3, vol. 5), p. 265.—Pfeif. Mon. Helic. vol. 5,
p. 183, and Mal. Blät, 1859, p. 24.
Rungtu Valley, near Darjiling.

PLATE LXXXVIII.

HELIX.

1, 4. **H. compluvialis**, Blanford, Journ. Asi. Soc.
Beng. 1865, pt. 2 (vol. 34), p. 66 : Cont. Mal. pt. 5.
Arracan Hills.

2, 3. **H. convallata**, Benson, An. Nat. Hist. 1856,
(ser. 2, vol. 18), p. 250.—Pfeif. Mon. Helic. vol. 4,
p. 46 : Novit. Conch. vol. 1, pl. 36, f. 14, 15, 16.
Birmah.

5, 6. **H. consepta**, Benson, An. Nat. Hist. 1860 (ser.
3, vol. 6), p. 190, and 1862, (ser. 3, vol. 11), p.
320.—Pfeif. Mon. Helic. vol. 5, p. 259.
Near Moulmein.

7, 10. **H. petrosa**, Hutton, Journ. Asi. Soc. Beng.
1834 (vol. 3), p. 84, for Helix no. 3 (from type).—
Pfeif. Mon. 1, p. 56.—H. vitrinoides, Pfeif. (not
Desh.) Mon. 1, p. 56.
Bengal.
Indian collectors term this fragile species H. vitri-
noides, and it suits part of the figures (pl. 110, f. 11,
12), which illustrate that shell in Kuster's "Chemnitz."

It is certainly not the true *vitrinoides* of Deshayes
figured without known locality in Guérin's "Magasin
de Zoologie,"

8, 9. **H. cycloidea**, Albers, Malak. Blätt. 1857
(vol. 4), p. 89, pl. 1, f. 1, 2, 3.—Pfeif. Mon. Helic.
vol. 4, p. 43.
Near Moulmein.

PLATE LXXXIX.

HELIX.

1, 2, 3. **H. umbrina**, Pfeiffer, Mon. Helic. vol. 4.
p. 49.—Reeve, Conch. Icon. Helix, f. 1345.
Ceylon.

4, 5, 6. **H. vilipensa**, Benson, An. Nat. Hist. 1853
(ser. 2, vol. 12), p. 93.—Pfeif. Mon. Helic. vol. 4.
p. 49.
Ceylon.

7, 10. **H. tenuicula**, Adams, Proc. Zool. Soc. 1868.
p. 14, pl. 4, f. 9 (as Macrochlamys).
Sattara, Bombay.

8, 9. **H. Potsaus**, Benson, An. Nat. Hist. 1859
(ser. 3, vol. 3), p. 388.—Pfeif. Mon. Helic. vol. 5.
p. 97.
Phie Than, Tenasserim.

PLATE XC.

HELIX.

1, 4. **H. lovicula**, Benson, An. Nat. Hist. 1859
(ser. 3, vol. 3), p. 391.—Pfeif. Mon. Helic. vol. 5.
p. 48.
Phie Than, Tenasserim.

2, 3. **H. causia**, Benson, An. Nat. Hist. 1859 (ser. 3.
vol. 3), p. 388.—Pfeif. Mon. Helic. vol. 5, p. 118.
Phie Than, Tenasserim.

5, 6. **H. mucosa**, W. and H. Blanford, Journ. Asi.
Soc. Beng. 1861 (vol. 30), p. 353, pl. 1, f. 11.—
Pfeif. Mon. Helic. vol. 5, p. 51.
Near Pykara, at Conoor Ghát and Sorgoor
Ghát, Nilgherries.

7, 8, 9. **H. pauxillula,** Benson, An. Nat. Hist. 1859
(ser. 3, vol. 3), p. 390.—Pfeif. Mon. Helic. vol. 5,
p. 119.

Thyet Myo.

This is not identical with the species so named by
Gould (Exped. Shells, p. 40).

10. **H. honesta,** Gould, (not Reeve) Proc. Bost. Soc.
N. H. vol. 2, p. 99.—Pfeif. Mon. Helic. vol. 4,
p. 63.

Tavoy : Aracan Hills, Pegu.

PLATE XCI.

ALYCÆUS.

1, 4. **A. Feddenianus,** Theobald, Journ. Asi. Soc.
Beng. 1876, vol. 39, pt. 2, p. 397, pl. 18, f. 4.
Shan States, and Upper Salwen.

2, 3. **A. amphora,** Benson, An. Nat. Hist. 1856,
ser. 2, vol. 17, p. 226.—Pfeif. Mon. Pneum. vol. 2,
p. 34; Novit. vol. 1, pl. 35, f. 15, 16.
Moulmein, and Tenasserim.

5, 6. **A. pyramidalis,** Benson. An. Nat. Hist. 1856,
ser. 2, vol. 17, p. 225.—Pfeif. Mon. Pneum. vol. 2,
p. 225 : Novit. Conch. vol. 1, pl. 35, f. 13, 14.
Theraboin Hill, Tenasserim, Birmah.

7, 10. **A. Andamaniæ,** Benson, An. Nat. Hist. 1861,
ser. 3, vol. 7, p. 28.—Pfeif. Mon. Pneum. vol. 3,
p. 47.
Port Blair, Andaman Islands.

8, 9. **A. urnula,** Benson, An. Nat. Hist. ser. 2,
vol. 11 (1853), p. 284.— Pfeif. Mon. Pneum. vol. 2,
p. 34.
Darjiling, Himalayah.

PLATE XCII.

ALYCÆUS.

1, 1. **A. stylifer,** Benson, An. Nat. Hist. ser. 2, vol.
19 (1857), p. 204.—Pfeif. Mon. Pneum. vol. 2, p.
37; Novit. vol. 1, pl. 35, f. 24-27.
Darjiling, Sikkim Himalayah.

2, 3. **A. prosectus,** Benson, An. Nat. Hist. ser. 2,
vol. 19 (1857), p. 203.—Pfeif. Mon. Pneum. vol.
2, p. 36: Novit. vol. 1, pl. 35, f. 21-23.
Teria Ghát, Khasia Hills.

5, 6. **A. physis,** Benson, An. Nat. Hist. ser. 3, vol.
3 (1859), p. 179.—Pfeif. Mon. Pneum. vol. 3,
p. 48.
Rungun Valley, near Darjiling.

7, 10. **A. Ingrami,** Blanford, Journ. Asi. Soc. Beng.
1862, vol. 31, p. 135.—Pfeif. Mon. Pneum. vol. 2,
p. 48.
Near Tongnop, Aracan.

8, 9. **A. umbonalis,** Benson, An. Nat. Hist. 1856,
ser. 2, vol. 17, p. 225.—Pfeif. Mon. Pneum. vol. 2,
p. 36: Novit. Conch. vol. 1, pl. 35, f. 18, 19, 20.
Akaouktoung, near Irawadi, Birmah.

PLATE XCIII.

ALYCÆUS.

1, 4. **A. bifrons,** Theobald, J. Asi. Soc. Beng. 1870,
vol. 39, pt. 2, p. 396, pl. 18, f. 1.
Shan States.

2, 3. **A. strangulatus,** Hutton, Mss. in Pfeif. Zeitschr.
Malak. 1846, p. 86.—Pfeif. Mon. Pneum. vol. 1,
p. 120; Kuster, ed. Chemn. Cyclost. pl. 17, f. 7, 8,
and pl. 38, f. 35.
Landour.

5, 6. **A. hebes,** Benson, An. Nat. Hist. ser. 2, vol.
19 (1857), p. 204.—Pfeif. Mon. Pneum. vol. 2, p.
37: Novit. pl. 35, f. 28-31.
Teria Ghát, Khasia Hills.

7. **A. gemmula,** Benson, An. Nat. Hist. ser. 3, vol.
3, (1859), p. 179.—Pfeif. Mon. Pneum. vol. 3,
p. 52.
Rungun Valley.

8, 9. **A. humilis,** Blanford, J. Asi. Soc. Beng. 1862,
vol. 31, p. 136; Cont. Mal. pt. 3.—Pfeif. Mon.
Pneum. vol. 3, p. 49.
Akaouktoung, on the banks of the Irawadi :
near Myanoung.

10. **A. armillatus,** Benson, An. Nat. Hist. 1856,
ser. 2, vol. 17, p. 227.—Pfeif. Mon. Pneum. vol. 2,
p. 37.
Thyet Myo, near river Irawadi, Birmah.

PLATE XCIV.

A L Y C Æ U S.

1, 2, 3. **A. politus,** Blanford, J. Asi. Soc. Beng.
1865, vol. 34, p. 83.
Phoungdo, near Cape Negrais, Aracan.

4, 7. **A. nitidus,** Blanford, J. Asi. Soc. Beng. 1862,
vol. 31, p. 141.—Pfeif. Mon. Pneum. vol. 3, p. 54.
Tongoop, Aracan.

5, 6. **A. Richthofeni,** Blanford, J. Asi. Soc. Beng.
1863, vol. 32, p. 324.
Moulmein.

8, 9, 10. **A. Avæ,** J. Asi. Soc. Beng. vol. 32 (1863),
p. 323.
Hills E. of Mandelay, and Ava.

PLATE XCV.

A L Y C Æ U S.

1, 4. **A. constrictus,** Benson, An. Nat. Hist. ser. 2,
vol. 8 (1851), p. 188 (as Cyclostoma); vol. 10,
p. 272.—Pfeif. Mon. Pneum. vol. 2, p. 35: Kuster
ed. Chemn. Cyclos. pl. 49, f. 24, 25.
Darjiling, Sikkim Himalayah.

2, 3. **A. bembex,** Benson, An. Nat. Hist. ser. 3,
vol. 3 (1859), p. 178.—Pfeif. Mon. Pneum. vol. 3,
p. 46.
Rungun Valley, near Darjiling, Himalayah.

4, 6. **A. otiphorus,** Benson, An. Nat. Hist. ser. 3,
vol. 3 (1859), p. 178.—Pfeif. Mon. Helic. vol. 3,
p. 46.
Pankabari; Rungun Valley, near Darjiling,
Himalayah.

7, 8, 9. **A. graphicus,** Blanford, J. Asi. Soc. Beng.
1862, vol. 31, p. 137.—Pfeif. Mon. Pneum. vol. 3,
p. 46.
Aracan Hills.

Our figure 7, represents the variety from the Shan
States, referred to by Theobald (J. Asi. Soc. Beng.
1870), vol. 39, pt. 2, p. 398, pl. 18, f. 3.

10, and pl. 97, f. 7. **A. margarita,** Theobald,
MSS.
Shan Provinces.

Although very near graphicus, differs in form,
sculpture, and colouring.

PLATE XCVI.
A L Y C Æ U S.

1, 4 **A. cucullatus,** Theobald, J. Asi. Soc. Beng.
1870, vol. 39, pt. 2, p. 396, pl. 18, f. 2.
Shan States.

2, 3. **A. polygonoma,** Blanford, J. Asi. Soc. Beng.
1862, vol. 31, p. 140.—Pfeif. Mon. Pneum. vol. 3,
p. 51.
Aracan Hills.

5, 6. **A. plectocheilus,** Benson, An. Nat. Hist.
ser. 3, vol. 3, (1859), p. 180.—Pfeif. Mon. Pneum.
vol. 3, p. 53.
Rungun Valley.

7, 10. **A. succineus,** Blanford, J. Asi. Soc. Beng.
1862, vol. 31, p. 139.—Pfeif. Mon. Pneum. vol. 3,
p. 50.
Aracan Hills.

8, 9. **A. Vulcani,** Blanford, J. Asi. Soc. Beng. 1863,
vol. 32, p. 323.—Pfeif. Mon. Pneum. vol. 3, p. 47.
Peak of Puppa in Ava.

PLATE XCVII.
ALYCÆUS.

1, 4. **A. cronulatus,** Benson, An. Nat. Hist. ser. 3,
vol. 3, (1859), p. 180.—Pfeif. Mon. Pneum. vol. 3,
p. .
Rungun Valley.

2, 3. **A. Theobaldi,** Blanford, J. Asi. Soc. Beng. 1862,
vol. 31, p. 142.—Pfeif. Mon. Pneum. vol. 3, p. 49.
Khasi Hills.

5, 6. **A. sculptilis,** Benson, An. Nat. Hist. 1856, ser. 2,
vol. 17, p. 226.—Pfeif. Mon. Pneum. vol. 2, p. 35.
Thyet Myo, near Irawadi, not far from
the boundary of British Birmah.

7. **A. margarita,** Theobald, see plate 95, f. 10.

8, 9, 10. **A. glaber,** Blanford, J. Asi. Soc. Beng.
1865, vol. 34, p. 84: Cont. Mal. pt. 5.
Akyab, Aracan Hills, S. of harbour.

PLATE XCVIII.
S T R E P T A X I S.
See previous plate, viii.*

1, 4. **S. Layardiana,** Benson, An. Nat. Hist. 1853
(ser. 2, vol. 12), p. 90.—Pfeif. Mon. Helic. vol. 4,
p. 332.
Ceylon.

* In Plate viii. the figures 5 and 10 were erroneously transposed.

2, 3. **S. Cingalensis**, Benson. An. Nat. Hist. 1853
(ser. 2, vol. 12), p. 91.—Pfeif. Mon. Helic. vol. 4,
p. 333.
  Ceylon.

5, 6. **S. Perroteti**, Petit, Revue Zool. Cuv. 1841,
p. 100 (as Helix).—Pfeif. Mon. Helic. vol. 3, p. 288,
Kust. ed. Chemn. pl. 143, f. 29, 30, 31.
  Nilgherries.

7. **S. solidula**, Stoliczka, J. Asi. Soc. Beng. 1871,
vol. 40, pt. 2, p. 166, pl. 7, f. 10.
  Near Moulmein.

8, 9, 10. **S. exacuta**, Gould, Proc. Bost. vol. 6
(1856); Otia Conch. p. 220.—Pfeif. Mal. Blät.
1856, (vol. 3) p. 258 ; Mon. Helic. vol. 4, p. 331.
  Mergui, Birmah.

The abrupt increase of the antepenult whorl, and the
smoothness of the base towards the mouth, easily distinguish the species from S. Sankeyi. The ribs,
moreover, are coarser, and a second, though minute
parietal lamella, surmounts the principal one.

PLATE XCIX.

PLANORBIS.

See previous plates, xxxix, xl.

1, 4. **P. compressus**, Hutton, J. Asi. Soc. Beng.
vol. 3, p. 91 (No. 3), 93.—Bens, J. Asi. Soc. Beng.
vol. 5, p. 713.—Martens, Malak. Blät. vol. 14,
p. 213.
  Ganges, &c.

The P. Tondanensis of Mousson is now regarded by
that writer as a synonym.

2, 3. **P. rotula**, Benson, An. Nat. Hist. 1850, ser. 2,
vol. 5, p. 351.
  Near Moradabad.

5, 6, 7. **P. hyptiocyclos**, Benson, An. Nat. Hist.
1865, ser. 3, vol. 11, p. 89, and Pfeif. Mon. Helic.
vol. 5, p. 117 (as Helix).
  Fort M'Donald, Ceylon.

Despite of its supposed habitat, there can be no
doubt as to the generic allocation. We have figured
Benson's own examples.

8, 9, 10. **P. convexiusculus**, Hutton, J. Asi. Soc.
Beng. vol. 18, pt. 2, (1849), p. 657.
  Affghanistan.

PLATE C.

PUPA (including **ENNEA**).

1. **P.** (En.) **Pirreei**, Pfeiffer, Proc. Zool. Soc. 1854,
p. 295 (as Pupa): Mon. Helic. vol. 4, p. 341 and
Novit. Conch. vol. 1, p. 199, pl. 32, f. 12, 13, 14
(as Ennea).
  Khoondah Mountains, near Calicut.

2. **P.** (En.) **Blanfordi**, Godwin-Austen, Proc. Zool.
Soc. 1872.
  Khasi Hills.

3. **P.** (En.) **vara**, Benson, An. Nat. Hist. 1859 (ser. 3,
vol. 3), p. 188 (as Pupa, sect. Ennea).—Pfeif. Mon.
Helic. vol. 5, p. 455 (as Ennea).
  Nandai, Khasi Hills.

4. **P.** (En.) **Ceylanica**, Pfeiffer, Proc. Zool. Soc.
1855, p. 9 (as Pupa, sec. Ennea).—Ennea C. Pfeif.
Mal. Blät. vol. 2 (1855) p. 63.—Novit. Conch.
vol. 1, pl. 32, f. 18, 19, 20; Mon. Helic. vol. 4,
p. 341.
  Ceylon.

5. **P.** (En.) **fartoides**, Theobald, J. Asi. Soc. Beng.
vol. 39, pt. 2 (1870), p. 400 (as Pupa).
  Shan Provinces.

6. **P.** (En.) **bicolor**, Hutton, J. Asi. Soc. Beng. 1834,
vol. 3, p. 86, 93 (as Pupa).—Pfeif. Mon. Helic.
vol. 2, p. 353, and Kust. ed. Chemn. Pupa, pl. 13,
f. 9, 10 (as Pupa).—P. nuellita, Gould, Proc. Bost.
1846, vol. 2, p. 99.—Pfeif. Mon. Hel. vol. 3, p. 545.
  Mirzapore, &c.; Tavoy, &c.: Ceylon.
For further references, see Pfeiffer, Mon. Helic.
vol. 6, p. 312.

7. **P. bathyodon**, Benson, An. Nat. Hist. 1863 (ser. 3,
vol. 11), p. 426.—Pfeif. Mon. Helic. vol. 6, p. 326.
  Nerbudda.

8. **P.{plicidens**, Benson, An. Nat. Hist. ser. 2, vol. 4
(1849, Aug.).—Pfeif. Mon. Helic. vol. 3, p. 353:
Kust. ed. Chemn. Pupa, pl. 17, f. 23, 24.
  Himalayah.

9. **P. salwiniana**, Theobald, J. Asi. Soc. Beng. 1870
(vol. 39, pt. 2), p. 400.
  Shan States.

Our figure is unsatisfactory: the body is usually
broader, and the mouth more oblique. The internal
lamellæ cannot, of course, be represented on so small
a scale: they average with the teeth and denticles
about nine in number.

10. **P. lapidaria**, Hutton, J. Asi. Soc. Beng. 1849,
vol. 18, pt. 2, p. 652.—Pfeif. Mon. Helic. vol. 4, p. 672.
  Affghanistan.

PLATE CI.

PUPA.

1. **P. stenopylis**, Benson, An. Nat. Hist. 1860 (ser. 3, vol. 5), p. 460.—Pfeif. Mon. Helic. vol. 5, p. 455.— G. Aust. Proc. Zool. 1872, pl. 30, f. 5. Darjiling.

2. **P. planguncula**, Benson, An. Nat. Hist. 1863 (ser. 3, vol. 11), p. 426.—Pfeif Mon. Helic. vol. 6, p. 359.
Orissa.
Our sole type having been smashed while in the artist's possession, we cannot vouch for the correctness of this drawing. If another should be procurable, and this drawing prove incorrect, the figure will be repeated.

3. **P. Huttoniana**, Benson, An. Nat. Hist. 1849 (ser. 2, vol. 4), p. 126.—Pfeif. Mon. Helic. vol. 4, p. 676. Simla.

4. **P. Himalayana**, Hutton, in Bens. An. Nat. Hist. 1863 (ser. 3, vol. 11), p. 428.—Pfeif. Mon. Helic. vol. 6, p. 299.
Simla and Mussoorie, W. Himalayah.

5, 6. **P. Evezardi**, Blanford, MSS.
Singhur Hill, Dekkan.

7. **P. eurina**, Benson, An. Nat. Hist. 1864 (ser. 3, vol. 13), p. 139.—Pfeif. Mon. Helic. vol. 6, p. 300.
At the river Gogra.

8. **P. seriola**, Benson, An. Nat. Hist. 1863 (ser. 3, vol. 11), p. 427.—Pfeif. Mon. Helic. vol. 6, p. 301.
Cuttack in Orissa.
Benson states that in one of the two types there is a rather distant tooth midway between the two lips, and that the pillar lip is broadly expanded at its commencement. The type having been smashed, and an imperfect photograph alone preserved, we cannot vouch for the correctness of our figure.

9. **P. gutta**, Benson, An. Nat. Hist. 1864 (ser. 3, vol. 13), p. 138.—Pfeif. Mon. Helic. vol. 6, p. 298.
Spiti Valley, Kunawar.
Both the known types have been crushed. Our drawing is taken from a photograph.

10. **P. diopsis**, Benson, An. Nat. Hist. 1863 (ser. 3, vol. 11), p. 427.—Pfeif. Mon. Helic. vol. 6, p. 506.
Valley of the Nerbudda.
We have figured the unique original, which does not well display the remote columellar tooth ascribed to it: the name, says Benson, was a misprint for diploos.

PLATE CII.

ACHATINA : Section Electra chiefly.

See previous plates xvii, xviii, xxxv, xxxvi, lxxviii.

1. **A. paupercula**, W. and H. Blanford, J. Asi. Soc. Beng. 1861, p. 382, pl. 1, f. 16: Cont. Mal. pt. 2.—Pfeif. Mon. Helic. vol. 6, p. 227.
Kolamullies, Shevroys, and Pachamullies.

2. **A. Deshayesiana**, Pfeiffer, Proc. Zool. Soc. 1852; Mon. Helic. vol. 3, p. 495: Kust. ed. Chemn. Bulimus, pl. 43, f. 13 to 16, Achat. No. 107.
Ceylon, and Koonalah Hills.

3. **A. Bensoniana**, Pfeiffer, Zeits. Malak. 1854, p. 27; Mon. Helic. vol. 3, p. 494; Kust. ed. Chemn Bulim. pl. 26, f. 12, 13.
Nilgherries.

4. **A. Punctogallana**, Pfeiffer, Mon. Helic. vol. 3, p. 493, for A. Ceylanica, Reeve (not Pf.) Conch. Icon. Achat. f. 59.
Ceylon.

5. **A. Mullorum**, Blanford, J. Asi. Soc. Beng. 1861, p. 362, pl. 1, f. 17.—Pfeif. Mon. Helic. vol. 6, p. 228.
Madras.

6. **A. Peguensis**, Blanford, J. Asi. Soc. Beng. vol. 34, 1865, p. 78.—Pfeif. Mon. Helic. vol. 6, p. 228.
Irawadi Valley, Pegu.

7. **A. rugata**, Blanford, J. Asi. Soc. Beng. 1870, vol. 39, pt. 2, p. 20, pl. 3, f. 18 (as Glessula).
Singhur, near Poona: a var. from Poorundhur.

8. **A. Beddomei**, Blanford, J. Asi. Soc. Beng. 1866, p. 41 : Cont. Mal. pt. 6.—Pfeif. Mon. Helic. vol. 6, p. 222.
Anamullay Hills.
Our figure is copied from an exquisite painting lent us by the author.

9. **A. illustris**, Godwin-Austen, MSS.
Cachar.

10. **A. balanus**, Benson, in Reeve, Conch. Icon. Achat. f. 109.—Pfeif. Mon. Helic. vol. 4, p. 627.
Near Agra : Kattiwar.

PLATE CIII.

A L Y C Æ U S.

See previous plates xci to xcvii.

1. **A. diagonus**, Godwin-Austen, J. Asi. Soc. Beng. 1871 (vol. 40, pt. 2), p. 88, pl. 3, f. 2. Diyung Valley, N. of Asálé, N. Cachar.

2, 3. **A. crenatus**, Godwin-Austen, J. Asi. Soc. Beng. 1871 (vol. 40, pt. 2), p. 91, pl. 3, f. 5. Burrail Range, N. Cachar.

4. **A. vestitus**, Blanford, J. Asi. Soc. Beng. 1862 (vol. 31), p. 138: Cont. Mal. pt. 3. Moditoung, Aracan Hills.

5, 6. **A. Khasiacus**, Godwin-Austen, J. Asi. Soc. Beng. 1871 (vol. 40, pt. 2), p. 90, pl. 3, f. 4. Khasi and Jaintia Hills.

7, 10. **A. pusillus**, Godwin-Austen, J. Asi. Soc. Beng. 1871 (vol. 40, pt. 2), p. 90, pl. 3, f. 3. Banks of Kopili River, from Jawai to Asálú.

8, 9. **A. conicus**, Godwin-Austen, J. Asi. Soc. Beng. 1871 (vol. 40, pt. 2), p. 87, pl. 3, f. 1. E. of Kopili River, N. Cachar.

PLATE CIV.

CYCLOPHORUS.

See previous plates i to iv, xxxiii, xxxiv, xlvii, xlviii.

1. **C. affinis**, var. picta, Theobald. A most lovely shell, which has sometimes minute spiral striolæ, sometimes concentric wrinkles.

2, 3. **C. Layardi**, H. Adams, Proc. Zool. Soc. 1868, p. 294, pl. 4, f. 21. Ceylon. We have only seen two individuals of this well-marked shell, both formerly possessed by H. Adams.

4. **C. speciosus**, Philippi, Zeits. Malak. 1847, p. 123 (as Cyclost.).—Pfeif. Mon. Pneum. vol. 1, p. 56, and in Kust. ed. Chemn. Cyclos. pl. 25, f. 1, 2, 3 (as Cyclostoma).—Reeve, Conch. Icon. Cyclop. f. 4. Birmah.

5, 6. **C. cornu-venatorium**, Sowerby, Thesaur. Conch. vol. 1, pl. 24, f. 41 (not 42), as Cyclost.—Benson, Ann. Nat. H. 1857 (ser. 2, vol. 19), March.—Pfeif. Mon. Pneum. vol. 2, p. 69 (as Cyclop. ?) partly. Ava, and Shan Hills.

In Sowerby's Monograph of Cyclostoma, this shell was confused with an Auloponma (f. 42), and erroneously identified with the Pterocyclos, figured by Chemnitz as "Cornu-venatorium" (the Helix. c. v. of Gmelin's Systema).

7. **C. speciosus**, var. Philippi. The lip of this well-marked form is peculiar for its colouring.

PLATE CV.

CYCLOPHORUS and LEPTOPOMA.

See previous plates i to iv, vi, xxxiii, xxxiv, xlvii, xlviii.

1. **L. conulus**, Pfeiffer, Proc. Zool. Soc., 1854, p. 127 (as Cyclostoma).—Reeve, Conch. Icon. Lept. f. 45. Ceylon.

2, 3. **C. scurra**, Benson, An. Nat. Hist. 1857 (ser. 2, vol. 19), p. 207.—Pfeif. Mon. Pneum. vol. 2, p. 47. Pegu.

4. **C. porphyriticus**, Benson, An. Nat. Hist. 1851 (ser. 2, vol. 8), p. 187 (as Cyclostoma).—Pfeif. in Kust. ed. Chemn. Cyclost. p. 383, pl. 50, f. 22 to 24 (as Cyclos.): Mon. Pneum. vol. 2, p. 49.—C. perdix, Reeve, Conch. Icon. Cyclop. f. 21. Eastern side of the Bay of Bengal. This is the shell which was at first referred by Benson (Zool. Journ.) to the C. perdix of Broderip and Sowerby (Zool. Journ.) a species from Tenasserim, with four whorls and a low spire, which does not well agree with the characters of C. perdix of Sowerby in his monograph (Thes. Conch. vol. 1), or of Pfeiffer in Kuster's edition of Chemnitz.

5, 6. **C. ravidus**, Benson, An. Nat. Hist. 1851 (ser. 2, vol. 8), p. 190 (as Cyclostoma).—Pfeif. Mon. Pneum. vol. 2, p. 67 : Kust. ed. Chemn. Cycl. pl. 50, f. 14, 15, 16 (as Cyclost.)—Reeve, Conch. Icon. Cyclop. f. 102. Nilgherries. All these figures and descriptions appear taken from the unique type we have again delineated.

7, 8. **C. stenostoma**, Sowerby, Thesaur. Conch. i, p. 95, pl. 31, f. 261 (as Cyclost.).—Pfeif. Mon. Pneum. vol. i, p. 9 ditto).—Reeve, Conch. Icon. Cycloph. f. 82. Figure 8 is taken from a paler variety which is less common than the other.

9. **C. stenostoma**, var. anguis, Sowerby.

Top of Nilgherries, with the preceding.

This shell (possibly a distinct species) is not merely larger than the preceding, but has the entire upper surface concentrically shagreened by wavy and somewhat broken raised wrinkles.

10. **C. cadiscus**, Benson, An. Nat. Hist. 1860, ser. 3, vol. 5, p. 385.—Pfeif. Mon. Pneum. vol. 3, p. 67. Matelle, Ceylon.

This is the same shell we had given another view of in Part I. (pl. 3, f. 8, not 9), as a form of Thwaitesii; at that time neither the collections of Benson nor Layard were adequately known to us, and even now the limits of parapsis, cadiscus, and Thwaitesii are not clear; Benson's ideal of the last, however, does not seem that of Pfeiffer.

PLATE CVI.

CATAULUS.

1. **C. Templemani**, Pfeiffer, Proc. Zool. Soc. 1852, p. 158 (as Pupina).—Mon. Pneum. vol. 2, p. 87.— Sow. Thes. Conch. vol. 3, pl. 264, f. 12. Ceylon.

2. **C. duplicatus**, Pfeiffer, Proc. Zool. Soc. 1851, p. 203.—Mon. Pneum. vol. 2, p. 90.—Sow. Thes. Conch. vol. 3, pl. 264, f. 2. Ceylon.
Near the last, but has less convex whorls.

3. **C. Blanfordi**, Dohrn, Proc. Zool. Soc. 1862, p. 202.—Pfeif. Mon. Pneum. vol. 3, p. 88 : Novit. Conch. vol. 1, pl. 58, f. 11, 12, 13.
Bollegalle in Ceylon.

4. **C. hæmastoma**, Pfeiffer, Proc. Zool. Soc. 1856, p. 331.—Mon. Pneum. vol. 2, p. 89.—Sow. Thes. Conch. vol. 3, pl. 264, f. 11. Ceylon.

5. **C. decorus**, Benson, An. Nat. Hist. 1853 (ser. 2, vol. 12, p. 96).—Sow. Thes. Conch. vol. 3, pl. 264, f. 13. Ceylon.

6. **C. Thwaitesii**, Proc. Zool. Soc. 1852, p. 147.— Mon. Pneum. vol. 1, p. 138.—Sow. Thes. Conch. vol. 3, pl. 264, f. 15. Ceylon.

---

7. **C. Austenianus**, Benson, An. Nat. Hist. 1853 (ser. 2, vol. 12), p. 95.—Pfeif. Mon. Pneum. vol. 2, p. 88.—Sow. Thes. Conch. vol. 3, pl. 264, f. 9. Henerat Jodde, Ceylon.

8. **C. Layardi**, Gray, MSS. in Pfeif. Proc. Zool. 1852, p. 160, pl. 13, f. 6.—Pfeif. Mon. Pneum. pt. 2, p. 87.—Sow. Thes. Conch. vol. 3, pl. 264, f. 7. Ceylon.

9. **C. aureus**, Pfeif. Proc. Zool. Soc. 1855, p. 118 : Mon. Pneum. vol. 2, p. 88. Ceylon.

10. **C. Calcadensis**, Blanford, J. Asi. Soc. Beng. 1869 (vol. 38, pt. 2), p. 157, pl. 16, f. 8.
Calcad Hills, beyond the limits of Travancore. Our example (named by the author) somewhat differs from the figure in the journal, but the species is said by Blanford to be very variable.

PLATE CVII.

UNIO.

See previous plates ix to xii, xli to xlv.

1. **U. Indicus**, Sowerby, in Reeve's Conch. Icon. Unio, f. 222.
Nerbudda River.
As yet so rare a shell, that we have seen only two adult specimens of the normal form.

2. **U. triembolus**, Benson, An. Nat. Hist. ser. 3, vol. 10 (Septen. 1862), p. 190.
Nerbudda River.

3. **U. favidens**, var. Benson. See Plate XI. f. 3.
Near the "Seven Tanks," Calcutta.
This dwarf form ranges between pinax and plagiosous. Our figured example is solid, and displays a strongly marked sculpture, but we possess others (which remind us of the unrecognised Rajahensis) that are almost smooth, and comparatively fragile.

4. **U. Indicus**, var. aurea, Sowerby.
Nerbudda River.
Possibly the link between Indicus and Triembolus.

5. **U. exolescens**, Gould, Proc. Nat. H. Boston, vol. 1, p. 141 : Otia Conch. p. 191.
Tavoy, Birmah.
Of the two examples received from Gould, the larger (a more beaked form) was in too bad a condition to be delineated ; the one we have figured is immature. The general aspect of the species reminds

us of the U. mutabilis of Lea (Obs. Unio, vol. 7), said to be Australian.

6, 7. **U. Sikkimensis**, Lea. See Plate XI. f. 4.
Assam.

This species, formerly so scarce, had been previously figured by us from an uncharacteristic example. The large variety (fig. 7) reminds one externally of Wynegungaensis.

PLATE CVIII.

PALUDOMUS.

1. **P. stomatodon**, Benson, An. Nat. Hist. 1862 (ser. 3, vol. 10), p. 415 (as ? Tanalia).
Near Cottyam, Travancore.

2. **P. rotunda**, Blanford, J. Asi. Soc. Beng. 1870 (vol. 39, pt. 2), p. 9, pl. 3, f. 2.
Travancore.

3. **P. lævis**, Layard, Proc. Zool. Soc. 1854, p. 89.
Ceylon.

4. **P. reticulata**, Blanford, J. Asi. Soc. Beng. 1870 (vol. 39, pt. 2), p. 9, pl. 3, f. 1.
Cachar.

5. **P. regulata**, Benson, An. Nat. Hist. 1856 (ser. 2, vol. 17), p. 496.
Thyet Mio, Birmah.

6. **P. regulata**, var. Benson.
Upper Birmah.

7. **P. parva**, Layard, Proc. Zool. Soc. 1854, p. 90.
Ceylon.

By some error the longitudinal brown marks (in the type) were described as spiral.

8. **P. ornata**, Benson, An. Nat. Hist. 1856 (ser. 2, vol. 17), p. 496.
Birmah.

9. **P. labiosa**, Benson, An. Nat. Hist. 1856 (ser. 2, vol. 17), p. 495.
Tongoop, Tenasserim, Birmah.

10. **P. monile**, Thorpe, MSS. Southern India.

PLATE CIX.

MELANIA.

See previous plates lxxi to lxxv, cix.

1. **M. spinata**, Godwin-Austen, Proc. Zool. Soc. 1872, p. 514, pl. 30, f. 1 (as Melanoides).
Kopili River, N. Cachar.

2. **M. variabilis**, (var. varicosa) Benson, J. Asi. Soc. Beng. vol. 5 (1836), p. 746 (the A. of his Gleanings in Science, vol. 1, pl. 7).—Voy. Bonite, p. 543, pl. 31, f. 13.
M. Carolinæ? Gray, Griffith's ed. Cuvier, Moll. pl. 13, f. 3 (1833).—M. varicosa, Troschel, Wiegm. Archiv. Nat. 1837, p. 174.—Philippi, Abbild. N. Conch. vol. 1, Mel. pl. 2, f. 2, 3.
River Gumti at Gospur, and Tolly's Nullah near Calcutta : River Hoogly, Calcutta.

3. **M. variabilis** (var. echinata), Benson.
Assam.

4. **M. fuscata**, Born, Ind. Test. Vindob. pl. 16, f. 17 (as Helix).—Dillwyn, Des. Cat. Shells, vol. 2, p. 951 (as Helix).—? Helix ater, Chemnitz, Conch. Cab. vol. 9, pt. 2, p. 116, f. 1229 (copied in Wood's Ind. Test. pl. 34, f. 144 (as H. fuscata).—Bulinus fuscus, Brug. Ency. Méth. Vers, vol. 1, p. 352.
Puniar (or Punmar) River, Cuddalore.

We have seen only two examples of this shell, which seems to have been delineated in Lister's Hist. Conch. pl. 116, f. 11 : they resemble the M. aspérans, var. of Reeve (C. Icon. Mel. f. 53, b.) but are not like the types of Hind's species.

5. **M. variabilis**, (var. cincta), Benson.
" Assam" (fide Bacon).

6. **M. variabilis**, (var. aspera), Benson. 543, pl. 31, f. 12, 15.
Hindostan.

7. **M. Herculea**, var. Sowerbii, Gould.
M. variabilis, Reeve, Conch. Syst. pl. 194, f. 2 (for the unnamed Melania of Sow. Genera Shells).—M. Herculea, Reeve, Conch. Icon. Melan. f. 4.
Tenasserim.

A remarkable form which approaches almost equally variabilis and Herculea.

PLATE CX.

MELANIA.

1. **M. tigrina**, Hutton, J. Asi. Soc. Beng. vol. 18, pt. 2 (1849), p. 658.
Affghanistan.

Compare this with the M. beryllina, of Brot, from Pondicherry (Rev. and Mag. Zool. 1860, pl. 17, f. 8).

2. **M. tigrina**, var. Hutton.
W. Himalayah.

3. **M. pyramis**, Benson, J. Asi. Soc. Beng. vol. 5
(1836), name for the Melania species B, in the
Gleon. Science Calcutta, vol. 2 (1830), p. 22.
River Goomty.
The shell figured is from the collection of Benson.
We presume not to assert that it is distinct from either
tuberculata or tigrina, but it is important to indicate the
exact type.

4. **M. pyramis**, var.—M. adspersa, Troschel, Wiegm.
Archiv. Naturg. 1837, p. 175, probably.
Shan States.
The M. adspersa of Philippi (Abbild. N. Conch.
vol. 3, p. 58, Melan. pl. 5, f. 5, 6) said by Brot to be
identical with the M. flammigera of the same work
(Melan. pl. 3, f. 11) does not equally suit our variable
species. Whether Philippi's specimens come from the
Ganges, as stated, may well be doubted.

5. **M. Hanleyi**, Godwin-Austen, Proc. Zool. Soc.
1872, p. 514, pl. 30, f. 2 (as Melanoides).
Diyung River, Cachar Hills.
Allied to the next, but the prickles are much more
numerous.

6. **M. Menkiana**, Lea, Obs. Unio, vol. 4, p. 24, for
M. plicata, Lea, Trans. Amer. Phil. Soc. (and Obs.
Unio, vol. 2, p. 20), pl. 23, f. 95 (not of Menke.
Synops. 1830).
Kheraaip, N. Cachar.
This rare shell may be easily distinguished from the
spinous forms of variabilis by the absence of those
coarse sulci which gird the base of the latter. It is not
the plicata of Reeve's figure, although two of the three
specimens in Cuming's collection are certainly Indian,
and not as stated from New Granada. It should be
noticed that Lea's figure was taken from a large
specimen with a cut-down lip, but all doubt as to its
identity is removed by the description. The M. spinosa
of Benson in Hanley's Conchological Miscellany (Mel.
pl. 1, f. 7) should rather have been referred to this
than to variabilis.

7 **M. scabra?** var. spinulosa.
Ceylon.
The shells represented in our figures 7 and 10, are
both called (but not described as) M. spinulosa by
Indian conchologists; yet neither can be positively
affiliated to the Lamarckian species from Timor delineated in Delessert's folio.

8, 9. **M. jugicostis**, Benson's MS.
Tenasserim River.

10. **M. acanthica** of Dohrn, Proc. Zool. Soc. 1858, as
of Lea (Proc. Zool. Soc. 1850, p. 194).
Ceylon.
The identity of this Melania (named from Dohrn's
type, now in the British Museum) with Lea's species
from the Philippine Islands, may possibly be questioned.

PLATE CXI.

HELIX.

See previous plates xiii to xvi, xxv to xxxii, l to
lxiv, lxxxiii to xc.

1. **H. Skinneri**, Reeve, Conch. Icon. Helix, f.
1387.—Pfeif. Mon. Helic. vol. 4, p. 219.
Ceylon.

2, 3. **H. undosa**, var. Blanford, J. Asi. Soc. Beng.
1865 (vol. 34), p. 68: Cont. Mal. pt. 5 (as
Nanina).
Shan Hills E. of Ava.
The original types (which we had not seen) were
much more shagreened, and less wrinkled than this
specimen.

4, 7. **H. ganoma**, Pfeiffer, Proc. Zool. Soc. 1855,
p. 124: Mon. Helic. vol. 4, p. 22.—Reeve, Conch.
Icon. Helix, f. 1267.—H. Juliana, Pfeif. in Kust
ed. Chem. Helix, pl. 33, f. 15.
Ceylon.
Very near the common and variable Juliana of Gray
(nearer of Sowerby's description in Beechey) to
which Pfeiffer preferentially refers the Dufourei of
Grateloup (changed from citrinoides, Grat.) in the
Act. Lin. Bordeaux, vol. 11, p. 107, pl. 1, f. 2.

5. **H. bajadera**, Pfeiffer, Mon. Helic. vol. 3, p. 52:
vol. 4, p. 250.—Reeve, Conch. Icon. Helix, f. 488.
Bengal.

6. **H. intumescens**, Blanford, J. Asi. Soc. Beng
1866 (vol. 36), p. 33: Cont. Mal. pt. 6 (as Nanina,
section Ariophanta)—Pfeif. Mon. Helic. vol. 5, p.
321.
Mahableshwar, W. Ghats of Hindostan.

PLATE CXII.

HELIX.

1, 2, 3. H. decussata, Benson, J. Asi. Soc. Beng.
1836 (vol. 5) p. 350 (as Nanina).—Pfeif. Mon.
Helic. vol. 1, p. 70.—Reeve, Conch. Icon. Helix, f.
743.

Bengal.

4, 5, 6. H. Sisparica, Blanford, J. Asi. Soc. Beng.
1866 (vol. 36), p. 34 (as Nanina).—Pfeif. Mon.
Helic. vol. 5, p. 122.

Sispara Ghat, Nilgherries.

Our figure is taken from a type lent us by Mr. W.
Blanford, to whose liberality and profound knowledge
of Indian Malacology the authors have been frequently
indebted.

7, 10. H. anserina, Theobald, J. Asi. Soc. Beng.
1865 (vol. 32, pt. 2), p. 4, name only.

Shan Provinces.

The conspicuously punctulate shagreen is an important character.

8, 9. H. Andersoni, Blanford, Proc. Zool. Soc.
1869, p. 446 (as Plectop.).

Bhamo, and Hoctone in Yunan.

PLATE CXIII.

AMPULLARIA.

1. A. cinerea, Reeve, Conch. Icon. Ampul. f. 94.
Ceylon.

The throat is usually chestnut, and there are
obscure bands under the epidermis as indicated by the
description.

2. A. corrugata, Swainson, Zool. Illust. ser. 1, pl.
120 (badly copied in Kuster's ed. Chemn. Ampul.
pl. 1, f. 10).

Bengal; Pondicherry, teste Belanger.

The only individual known to us agrees fairly with
the drawing of Swainson, who having cited, as a
synonym, the A. rugosa of Sowerby's genera (which
looks more like globosa) induced Deshayes to identify
both with his A. sphærica (Encycl. Méth. Vers). Can
it be an abnormal form of the next species?

3. A. globosa, Swainson, Zool. Illust. ser. 1, pl.
119.—Philippi, Monog. Ampul. (Kust. ed. Chemn.)
p. 8, pl. 1, f. 3.—Reeve, Conch. Icon. Amp. f. 46,
47.

Calcutta; Rohilkhund; Orissa, &c.

A common shell which attains much larger dimen-
sions than here exhibited: occasionally its peritreme is
tinted with orange red, so as to remind one strongly of
Crouch's ideal of A. Guyanensis (Crouch, Lam. pl.
15, f. 18). The A. rotundata of Say (erroneously
described as American, but with a testaceous operculum) is supposed to be a form of this species.

4. A. globosa, var. sphærica, from Moradabad.

5. A. globosa, var. fasciata, from Moradabad.

PLATE CXIV.

AMPULLARIA.

1. A. carinata, Swainson, Zool. Ill. ser. 2, from
which Philip. Mon. Ampul. (in Kust. ed. Chemn.)
pl. 1, f. 2—? Reeve, Conch. Icon. Ampul. f. 58.

Ceylon.

This is not the species so named by Lamarck (as the
Cyclostoma carinatum of Olivier) which belongs to the
sinistral genus Lanistes. It is chiefly distinguishable
from the next species by the greater breadth of the
penult whorl, and the sharper sutural angulation.

2. A. Malabarica, Philippi (not Reeve), Mon.
Ampul. (Kust. ed. Chemn.) p. 39, pl. 7, f. 8.

Cochin, Malabar; Bombay.

Very closely allied to the preceding, and not im-
probably a local variety. Yet the upper whorls of the
spire (which is more exserted) are rounded. The
young type of Reeve's Malabarica (f. 67) does not
exhibit the flat infrasutural ledge referred to by
Philippi. The A. pallens of Philippi (Kust. ed. Chemn.
Ampul. p. 52) may possibly prove the variety we have
received from Cashmire. Reeve's A. canaliculata, said
to have been taken in Cashmire, seems the young of
A. speciosa.

3. A. Tischbeini, Dohrn, Proc. Zool. Soc. 1858.
Ceylon.

Copied from the type in the British Museum.

4. A. Layardi, Reeve, Conch. Icon. Ampul. f. 27,
40.

Ceylon.

Our specimen bears more resemblance to the characteristic variety f. 40, than to the less strikingly distinctive figure 27 in Reeve's Iconica.

5. **A. Paludinoides**, Philippi, Kust. ed. Chemn. Ampul. p. 27, pl. 7, f. 4 (as of De Cristofori and Jan).—Reeve, Conch. Icon. Ampul. f. 9.
Mangalore ; near Moulmein ; Pegu.

Jan's species should be ignored, for it is so inadequately defined, that his description would suit half a score of Ampullariæ : it is not likely to be the shell here figured, as it is called umbilicated and South American.

6, **7. A. Paludinoides**, var.

A peculiar banded form from Pegu. Figure 7 reminds us of the Reevean (Conch. Icon. Ampul. f. 10) ideal of A. conica, a Singapore shell, which Mr. Hanly carefully identified with the young original type, and figured in his Conchological Miscellany. Von Martens states that Reeve's ideal is not that of Pfeiffer in his Novitates.

PLATE CXV.

AMPULLARIA, and PALUDINA.

For Paludinæ see previous plates lxxvi, lxxvii.

1. **A. nux**, Reeve, Conch. Icon. Ampul. f. 132.
Small streams from Bore Ghat ; Bombay ; Dekkan.

2. **A. Theobaldi**, Hanley.
Birmah ? or Pegu ?
This magnificent shell was given to Mr. Hanley by his conductor, but the precise locality was mislaid.

3. **A. Saxea**, Reeve, Conch. Icon. Ampul. f. 108.
Bassein, Pegu.

4. **A. Saxea**, var. Reeve.
Pegu.
Our specimen is selected from its extreme dissimilarity to the preceding, yet the spire is occasionally even still more depressed.

5. **A. Woodwardi**, Dohrn, Proc. Zool. Soc. 1858.
Ceylon.
The only two individuals known to us are in the British Museum.

6. **A. mœsta**, Reeve, Conch. Icon. Ampull. f. 92.
Ceylon.
Our figure was taken from the Cumingian type, now in the British Museum.

7. **P. digona**, Blanford, Proc. Zool. Soc. 1869, p. 445.
River Irawaddy, Birmah.
The individual here figured exhibits the distinctive features in a remarkable degree. The author alludes to the possibility of his shell being a variety of dissimilis ; its affinities seem rather zonata, filosa and fucolata.

8. **P. variata**, Frauenfeld, Verhandl. Zool. Bot. Wien, 1862, p. 1163 (as Vivipara v.).
The individual figured was the original Pondicherry specimen named by the author : Reeve's (Conch. Icon. Palud. f. 58) seems either a squat form of dissimilis or Bengalii.

9. **P. Ceylanica**, Dohrn, var. ecarinata.
Common in Ceylon.

PLATE CXVI.

TANYSIPHON, SCAPHULA, NOVACULINA.

1, 4. **T. rivalis**, Benson, An. Nat. Hist. 1858 (ser. 3, vol. 1), p. 407, pl. 12, B. f. 1 to 3.
From mud at low-water in streams near Calcutta.

2, 3. **S. Deltæ**, Blanford, J. Asi. Soc. Beng. vol. 36, pt. 2, pl. 14, f. 7–10 : Cont. Mal. pt. 8, p. 21, pl. 3, f. 7–10.
Banks of the Irawaddy, Pegu.

5, 6. **S. pinna**, Benson, An. Nat. Hist. 1856 (ser. 3, vol. 17), p. 128.—Blanf. J. Asi. Soc. Beng. vol. 36, pt. 2, pl. 14, f. 11–13.
Tenasserim River.

7. **N. Gangetica**, Benson, Glean. Science Calcut. vol. 2 (1830, Feb.), p. 63 (as genus Novaculina) : Ann. Nat. Hist. 1858 (ser. 3, vol. 1), pl. 12, B. f. 4.
River Jumna at Humeerpore, Bundelkhund.

8, 9. **S. celox**, Benson, J. Asi. Soc. Beng. vol. 5 (1836), p. 759 (as figured in Glean. Sc. Calcutta), vol. 1, pl. 7, f. 2, 3 : An. Nat. Hist. 1856, p. 129.—Blanf. J. Asi. Soc. Beng. vol. 36, pt. 2, pl. 14, f. 14, 15.
River Jumna, near Bundelkhund, &c.

The generic appellation of Scaphula was proposed by Benson for this shell, in the fifth volume of the Zoological Journal (1834) without any specific denomination.

10. N. Gangetica? var. Theobaldi, Benson.
Tenasserim River: Pegu.
This was regarded by Benson as a large variety of
his Gangetic species: it looks distinct, but without more
specimens its separation would, perchance, be un-
advisable.

9. O. macrostoma, Beddome's MS. in Blanf. J. Asi.
Soc. Beng. 1869 (vol. 38), p. 139, pl. 16, f. 7.
Bramagiri Hills, Wynaad, not far from the
Malabar Coast.

10. O. Nilgiricum, W. and H. Blanford, J. Asi. Soc.
Beng. 1861 (vol. 29), p. 121; Cont. Mal. Ind. pt.
1, p. 5: Proc. Zool. Soc. 1866, pl. 18, f. 13.
Pykara, top of Nilgherries, in fallen leaves.

PLATE CXVII.

ACMELLA, HYDROCÆNA, OPISTHOSTOMA.

1. A. torsa, Benson, An. Nat. Hist 1853 (ser. 2, vol.
11), p. 285 (Cyclostoma).—Pfeif. Mon. Pneum. vol.
2, p. 158, (as Hydrocæna).—Acicula t. Blanf. An.
Nat. Hist. 1869 (ser. 4, vol. 3), pl. 16, f. 2.
In moss from Khasi Hills.

2. H. Blanfordiana, Stoliczka, J. Asi. Soc. Beng.
1873 (vol. 41, pl. 2), p. 332, pl. 11, f. 5, 6.
Ataran Valley, near Moulmein.

3. H. pyxis, Benson, An. Nat. Hist. 1856 (ser. 2,
vol. 17), p. 232.—Pfeif. Mon. Pneum. vol. 2, p. 161.
Thyet Myo, Birmah: near Henzada, Pegu.

4. H. illex, Benson, An. Nat. Hist. 1856 (ser. 2, vol.
17), p. 231.—Pfeif. Mon. Pneum. vol. 2, p. 161.
Thyet Myo: Phie Than, Tenasserim.

5. H. frustillum, Benson, An. Nat. Hist. 1860 (ser.
3, vol. 6), p. 193.—Pfeif. Mon. Pneum. vol. 3, p.
251 (as Georissa).
Ava.

6. H. Rawesiana, Benson, An. Nat. Hist. 1860 (ser.
3, vol. 6), p. 193.—Pfeif. Mon. Pneum. vol. 3, p.
252 (as Georissa).—Th. and Stol. J. Asi. Soc. Beng.
1872 (vol. 41, pt. 2), p. 332 (as Georissa).
Near Moulmein.

7. H. saritta, Benson, An. Nat. Hist. 1851 (ser. 2,
vol. 8), p.188 (as Cyclostoma).—Pfeif. Mon. Pneum.
vol. 1, p. 314 (ditto).
Near Cherra Poongee, Garo Hills, beyond
Eastern boundaries of Bengal.

8. O. Fairbankii, Blanford, Proc. Zool. Soc. 1866,
p. 448, pl. 38, f. 14.
Near Khandalla, Western Ghats, between
Bombay and Poonah.

PLATE CXVIII.

CLAUSILIA.

See previous plate, xxiv.

1. C. Ceylanica, Benson, An. Nat. Hist. 1863
(ser. 3, vol. 11), p. 89.—Pfeif. Mon. Helic. vol. 6,
p. 427.—Blanf. J. Asi. Soc. Beng. 1872 (vol. 41,
pt. 2), pl. 9, f. 4.
Southern parts of Ceylon.
The whorls are densely set with longitudinal raised
wrinkles, which are spirally decussated on the final
volution.

2, 3. C. Gouldiana, Pfeiffer, Mal. Blät. vol. 3
(1856), p. 259: Mon. Helic. vol. 4, p. 724 : Nova.
Conch. vol. 1, pl. 34, f. 18 to 20.
Mergui: near Moulmein, British Birmah.
This may be the insignis of Gould (not of Pfeiffer).

4. C. ovata, Blanford, J. Asi. Soc. Beng. 1872
(vol. 41, pt. 2), p. 206, pl. 9, f. 17.
Toughu, Birmah.

5, 6. C. ferruginea, Blanford, J. Asi. Soc. Beng.
1872 (vol. 41, pt. 2), p. 202, pl. 9, f. 7.
Naga Hills, and North Cachar.

7. C. monticola, Godwin-Austen, in Blanford's
monograph in the J. Asi. Soc. Beng. 1872 (vol. 41,
pt. 2), p. 205, pl. 9, f. 13.
North Cachar.

8, 9. C. Arakana, Theobald, in Stolic. J. Asi. Soc.
Beng. 1872 (vol. 41, pt. 2), p. 210, pl. 9, f. 20.
Arracan Hills; Mai-i, Sandoway.

10. C. Philippiana, Pfeiffer, Zeits. Mal. 1847,
p. 69: Mon. Helic. vol. 2, p. 43: Kuster's ed.
Chemn. Claus. p. 100, pl. 11, f. 7, 8, 9.—Blanf. J.
Asi. Soc. Beng. 1872 (vol. 41, pt. 2), pl. 9, f. 14.
Moulmein, and near Mergui.

PLATE CXIX.

DIPLOMMATINA.

1, 4. **D. Austeni**, Blanford, J. Asi. Soc. Beng. 1868,
p. 81, pl. 3, f. 2 : Cont. Mal. pt. 9.
Cherra Poonji, and Moothericksan, in Khasi
Hills.

2, 3. **D. oligopleuris**, Blanford, J. Asi. Soc. Beng.
1868, p. 82, pl. 3, f. 4.
Kamuk Hill, Arracan : Teria Ghat, S. side of
Khasi Hills.

5, 6. **D. Blanfordi**, Benson, An. Nat. Hist. 1860
(ser. 3, vol. 5), p. 460.—Pfeif. Mon. Pneum. vol. 3,
p. 9.
Darjiling.

7. **D. pullula**, Benson, An. Nat. Hist. 1859 (ser. 3,
vol. 3), p. 182.—Pfeif. Mon. Pneum. vol. 3, p. 9.
Rungung on the West of Darjiling.

8. **D. Sherfaiensis**, Godwin-Austen, J. Asi. Soc.
Beng. 1870, p. 3, pl. 1, f. 3.
Khasi Hills.

9. **D. labiosa**, Blanford, J. Asi. Soc. Beng. 1868,
p. 80, pl. 2, f. 3 : Cont. Mal. pt. 9.
Khasi and Garo Hills.

10. **D. exilis**, Blanford, J. Asi. Soc. Beng. 1862
(vol. 32), p. 325.—Pfeif. Mon. Pneum. vol. 3, p. 10.
Mya Leit Doung, Ava, Burmah.

PLATE CXX.

DIPLOMMATINA.

1, 4. **D. gibbosa**, Blanford, J. Asi. Soc. Beng. 1868,
p. 80, pl. 2, f. 1 : Cont. pt. 9.
Habiang in Garo Hills.

2, 3. **D. Jaintiaca**, Godwin-Austen, J. Asi. Soc.
Beng. 1870, p. 4 (as the nov. spec. of vol. 37, pt. 2,
pl. 3, f. 3).
Khasi Hills.

5, 6. **D. depressa**, Godwin-Austen, J. Asi. Soc.
Beng. 1870, p. 2, pl. 1, f. 2.
Khasi Hills.

7. **D. semisculpta**, Blanford, J. Asi. Soc. Beng.
1868, p. 78, pl. 1, f. 6 : Cont. Mal. pt. 9.
Darjiling.

8, 9. **D. costulata**, Hutton MSS. (as Carychium) in
Benson, An. Nat. Hist. 1849 (ser. 2, vol. 4),
p. 191.—Pfeif. Mon. Pneum. vol. 1, p. 122.
W. Subhimalayah ; Landour.

10. **D. ungulata**, H. Blanford, J. Asi. Soc. Beng. 1871,
vol. 40, pt. 2, p. 42, pl. 2, f. 7.
Darjiling.

PLATE CXXI.

PALUDOMUS.

See previous plate, cviii.

1. **P. crinaceus**, Reeve, Proc. Zool. Soc. 1852, p.
128.—Tanalia c. Layard, Proc. Zool. 1854, p. 91.
Ceylon.

2. **P. loricata**, Reeve, Conch. Icon. vol. 1, Palud. f.
1, b, c.—Tanalia c. Layard, Proc. Zool. Soc. 1854,
p. 91.
In rapids flowing from Adam's Peak and
Calloo ganga, above Ratnapoora, Ceylon.

3. **P. undata**, Reeve, Conch. Icon. Palud. f. 2.
In rapids flowing from Adam's Peak, Ceylon.

4. **P. Skinneri**, Dohrn, Proc. Zool. Soc. 1857, p.
124.
Ceylon.

5. **P. orea**, Reeve, Proc. Zool. Soc. 1852, p. 128.—
Tanalia orea, Layard, Proc. Zool. Soc. 1854.
Mountain streams of Ceylon.

6. **P. Layardi**, Reeve, Proc. Zool. Soc. 1852, p.
128.
Mountain streams of Ceylon.

7. **P. Reevei**, Layard, Proc. Zool. Soc. 1854 (as
Tanalia), p. 92.
The Calloo ganga, Ratnapoora, Ceylon.

8, 9. **P. melanostoma**, Thorpe MSS.
Ceylon.

10. **P. regalis**, Layard, Proc. Zool. Soc. 1854, p. 93
(as Philopotamis).
Stream in the Cnin Corle, Western province,
Ceylon.

PLATE CXXII.

PALUDOMUS.

1. **P. similis**, Layard, Proc. Zool. Soc 1854 (as
Tanalia) p. 92.
A mountain torrent at Kandangasoa, near
Ratnapoora, Ceylon.

2. **P. sulcata**, Reeve, Conch. Icon. Palud. f. 8.
   In a mountain stream at Ratnapoora, Ceylon.

3. **P. distinguenda**, Dohrn, Proc. Zool. Soc. 1857, p. 124.
   Ceylon.

4. **P. solida**, Dohrn, Proc. Zool. Soc. 1857, p. 124.
   Ceylon.

5. **P. Tennentii**, Reeve, Conch. Icon. Palud. f. 12.
   In a rocky stream flowing from Adam's Peak, Ceylon.

6. **P. Gardneri**, Reeve, Conch. Icon. Palud. f. 9.
   In a stream at the foot of Adam's Peak, Ceylon.

7. **P. picta**, Reeve, Conch. Icon. Palud. f. 10.
   In a mountain stream at Ratnapoora, Ceylon.

8. **P. Neritoides**, Reeve, Conch. Icon. Palud. f. 3.
   In the bed of a river at Ambegamoa, Ceylon.

9. **P. dromedarius**, Dohrn, Proc. Zool. Soc. 1857, p. 124.
   Ceylon.

10. **P. stephanus**, Benson, J. Asi. Soc. Beng. 1836, vol. 5, p. 747 (as Melania).—Reeve, Conch. Icon. Palud. f. 11.—Mel. coronata, Von den Busch, in Philip. N. Conch. vol. 1, Mel. pl. 1, f. 5, 6. Bengal.

PLATE CXXIII.
PALUDOMUS.

1. **P. fulgurata**, Dohrn, Proc. Zool. Soc. 1857, p. 123.
   Ceylon.
   This species resembles the P. phasianina of the Seychelles.

2. **P. Chilinoides**, Reeve, Conch. Icon. Palud. f. 7.
   Ceylon.

3. **P. decussata**, Reeve, Proc. Zool. Soc. 1852, p. 127.
   Ceylon.

4. **P. clavata**, Reeve, Proc. Zool. Soc. 1852, p. 129.
   Mountain streams of Ceylon.

5. **P. globulosa**, Gray, Grif. ed. Cuvier, Mol. pl. 14, f. 6 (as Melania).—Reeve, Conch. Icon. Palud. f. 4.
   Ambegamoa, Ceylon.

6. **P. lutosa**, Souleyet, Voy. Bonite, Zool. p. 350, pl. 31, f. 28, 29, 30.
   Ganges.

7. **P. acuta**, Reeve, Proc. Zool. Soc. 1852, p. 127.
   Near Pondicherry.
   This and the preceding may prove mere forms of the next species.

8. **P. Tanschaurica**, Gmelin, Syst. Nat. 3655, for the Helix fluviatilis Tanschaurensis of Chemn. Conch. Cab. vol. 9, p. 174, f. 1243.—Helix fluviatilis, Dillwyn, Des. Cat. Shells, p. 959.
   Southern India (Coromandel, &c.).

9. **P. Paludinoides**, Reeve, Proc. Zool. Soc. 1852, p. 127.
   Sikkim branch of the Sylhet.

10. **P. bicincta**, Reeve, Proc. Zool. Soc. 1852, p. 129.
    Mountain streams of Ceylon.

PLATE CXXIV.
PALUDOMUS.

1. **P. nigricans**, Reeve, Conch. Icon. Palud. f. 6.
   In mountain streams at Ceylon.

2, 3. **P. Maurus**, Reeve, Proc. Zool. Soc. 1852, p. 127.
   Ganges.

4. **P. conica**, Gray, Grif. ed. Cuvier, Mol. pl. 14, f. 5 (as Melania).—Reeve, Conch. Icon. Palud. f. 14.—Benson, J. Asi. Soc. Beng. vol. 5, p. 747.—Mel. crassa, Busch, in Phil. N. Conch. vol. 1, Mel. pl. 1, f. 10, 11.—P. rudis, Reeve, Proc. Zool. 1852.
   Sylhet, Bhootan, Assam, &c.

5. **P. Reevei**, Layard, Proc. Zool. Soc. 1854 (as Tanalia), p. 92.
   The Calloo ganga, Ratnapoora, Ceylon.

6. **P. Swainsoni**, Dohrn, Proc. Zool. Soc. 1857, p. 125.
   Ceylon.

7. **P. nasuta**, Dohrn, Proc. Zool. Soc. 1857, p. 123.
   Ceylon.
   Figured from the sole type in the British Museum.

8. **P. sphaerica**, Dohrn, Proc. Zool. Soc. 1857, p. 124.
   Ceylon.
   Figured from the unique type described.

9. **P. torrenticola**, Dohrn, Proc. Zool. Soc. 1858.
   Ceylon.
   Figured from the type in the British Museum.

10. **P. baccula**, Reeve, Proc. Zool. Soc. 1852, p. 128.—Hanley, Conch. Misc. Melan. f. 63.
    Branch of the Ganges.

PLATE CXXV.

PALUDOMUS.

1, 4. **P. funiculata**, Reeve, Conch. Icon. Palud. f. 13.—Tanalia, f. Layard, Proc. Zool. Soc. 1854, p. 93 (amended description).

In a mountain stream, not far from Ratnapoora, Ceylon.

2, 3. **P. pyriformis**, Dohrn, Proc. Zool. Soc. 1858, p. 536.

Ceylon.

5, 6. **P. dilatata**, Reeve, Proc. Zool. Soc. 1852, p. 128.

Ceylon.

7. **P. abbreviata**, Reeve, Proc. Zool. Soc. 1852, p. 127.

Ceylon.

Forms too near an approach to bicinctus.

8, 9. **P. Thwaitesii**, Layard, Proc. Zool. Soc. 1854, p. 92 (as Philopotamis).

Ceylon.

10. **P. Hanleyi**, Dohrn, Proc. Zool. Soc. 1858, p. 535.

Ceylon.

The spiral lines are too minute to be adequately represented. All the Paludomi of this plate have been drawn from the specimens originally described from Cuming's collection (now in the British Museum).

PLATE CXXVI.

PALUDOMUS.

1, 4. **P. constricta**, Reeve, Proc. Zool. Soc. 1852, p. 129.

Ceylon.

A rather uncertain species, figured, as indeed are many (f. 5, 6, 8, 9,) in this plate, from the Cumingian types.

2, 3. **P. palustris**, Layard, Proc. Zool. Soc. 1854, p. 89.

Ceylon.

Remarkable for its granulated surface.

5, 6. **P. Cumingiana**, Dohrn, Proc. Zool. Soc. 1857, p. 124.

Ceylon.

Distinguishable from Gardneri, says the author, by its larger mouth, and the deep channel-like impression on the upper part of the whorl.

7, 10. **P. obesa**, Philippi, Abbild. Neue Conch. vol. 2, p. 170, Melania, pl. 4, f. 3 (as ?Melania). —P. maculatus, Lea, Proc. Nat. Philadel. vol. 8, p. 110.—Rivulina m. Lea, Journ. Acad. Philad. ser. 2, vol. 6, p. 118, pl. 22, f. 10.

Bombay : Ahmednuggur.

8, 9. **P. nodulosa**, Dohrn, Proc. Zool. Soc. 1857, p. 125.

Ceylon.

PLATE CXXVII.

HELIX.

See previous plates xiii to xvi, xxv to xxxii, l to lxiv, lxxxiii to xci, cxi, cxii.

1. **H. Waltoni**, Reeve, Proc. Zool. Soc. 1842, p. 49 : Conch. Syst. vol. 2, pl. 166, f. 2, 3 : Conch. Icon. Helix. f. 372.—Pfeif. Mon. Helic. vol. 1, p. 19.

Ceylon.

The ordinary form has a speckled epidermis.

2. **H. hæmastoma**, Linnæus, Syst. Nat. ed. 12, p. 1247.—Chemn. Conch. Cab. vol. 9, f. 1150,1.—Reeve, Conch. Icon. Helix, f. 366, b.—Pfeif. Mon. Helic. vol. 4, p. 195.

Ceylon.

There is a black-mouthed variety very different from the true melanotragus.

3. **H. melanotragus**, Born (not Reeve), Test. Mus. Vind. p. 388.—H. hæmastoma, var., Chemn. Conch. Cab. vol. 9, f. 1052, 3, and Reeve, Conch. Ic. Helix, f. 366, c.

Ceylon.

Born referred solely to Geve (f. 329) for an illustration (and it is a characteristic one) of his species, which, whether really distinct, or merely a varietal form of the last, may usually be recognised by its broad white upper band, and the shape of its earlier volutions.

4. **H. superba**, Pfeiffer, Zeits. Mal. 1850, p. 71 : Mon. Helic. vol. 3, p. 185 : Kust. ed. Chemn. Helix, pl. 133, f. 1, 2.—Reeve, Conch. Icon. Helix, f. 368.

Ceylon.

5. **H. fastosa**, Albers, Mal. Blät. 1854 (vol. 1), p. 213.—Pfeif. Mon. Helic. vol. 4, p. 197.

Ceylon.

6. H. phœnix, Pfeiffer, Mon. Helic. vol. 4, p. 194.
—H. melanotragus, var. Reeve, Conch. Ic. Helix, f. 367, b.
Ceylon.

7. H. Grevillei, Pfeiffer, Proc. Zool. Soc. 1856, p. 387, pl. 36, f. 8 : Mon. Helic. vol. 4, p. 195.
Ceylon.
The oblique folds are not constant. The shell is a link between phœnix and superba.

PLATE CXXVIII.
HELIX.

All the figures are somewhat enlarged, and are drawn from the Cumingian types in the British Museum.

1, 4. H. acallos, Pfeiffer, Proc. Zool. Soc. 1856, p. 327 : Mon. Helic. vol. 4, p. 94.
Nilgherries.

2, 3. H. carneola, Pfeiffer, Proc. Zool. Soc. 1854, p. 148 : Mon. Helic. vol. 4, p. 47.—Reeve, Conch. icon. Helix, f. 1374.
Ceylon.

5, 6. H. convexiuscula, Pfeiffer, Proc. Zool. Soc. 1855, p. 191 : Mon. Helic. vol. 4, p. 35.
Ceylon.

7, 10. H. Thwaitesi, Pfeiffer, Proc. Zool. Soc. 1853, p. 125 : Mon. Helic. vol. 4, p. 50.—Reeve, Conch. Icon. Helix, f. 1336.
Ceylon.

8, 9. H. vallicola, Pfeiffer, Proc. Zool. Soc. 1854, p. 289 : Mon. Helic. vol. 4, p. 46.—Reeve, Conch. Icon. Helix, f. 1397.
Koondah Mountains, near Calicut.

PLATE CXXIX.
HELIX.

All the figures are greatly enlarged.

1, 2, 3. H. cuomphalos, W. and H. Blanford, J. Asi. Soc. Beng. 1861 (vol. 30), p. 354 : Cont. Mal. pt. 2, p. 8.—Pf. Mon. Helic. vol. 5, p. 138.
Near Pykara, Nilgherries.

4. H. febrilis, W. and H. Blanford, J. Asi. Soc. Beng. 1861 (vol. 30), p. 357, pl. 2, f. 4 : Cont. Mal. pt. 2.—Pfeif. Mon. Helic. vol. 5, p. 76.
Kalryennullies, Southern India.

5, 6. H. conulus, Blanford, J. Asi. Soc. Beng. 1865, pt. 2 (vol. 34), p. 73, and Cont. Mal. pt. 5, p. 9 (as Nanina).—Pf. Mon. Helic. vol. 5, p. 89.
Arracan.

7, 10. H. tricarinata, W. and H. Blanford, J. Asi. Soc. Beng. 1861 (vol. 30), p. 355, pl. 1, f. 10 : Cont. Mal. pt. 2, p. 9.—Pfeif. Mon. Helic. vol. 5, p. 91.
Near Pykara, Nilgherries.

8, 9. H. miccyla, Benson, Ann. Nat. Hist. 1860 (ser. 3, vol. 5), p. 384.—Pfeif. Mon. Helic. vol. 5, p. 55.
Matelle, Ceylon.

PLATE CXXX.
HELIX.

1, 4. H. Bactriana, Hutton, Journ. Asi. Soc. Beng. 1849 (vol. 17, pt. 2), p. 65 : Pfeif. Mon. Helic. vol. 4, p. 128.—Reeve, Conch. Icon. Helix. f. 1376.
Candahar.

2, 3. H. vidua, Blanford, MSS.
Khasi Hills.

5, 6, 7. H. Patane, Benson, Ann. Nat. Hist. 1859 (ser. 3, vol. 3), p. 270 : Pfeif. Mon. Helic. vol. 5, p. 11 3 : Mal. Blät. 1859, p. 22.
Darjiling.

8, 9. The only example known to Mr. Hanley in England having been utterly smashed, he does not dare to identify the species. Its sculpture was remarkable.

10. H. vittata, Müller, Hist. Verm. pt. 2, p. 75.—Reeve, Conch. Icon. Helix, f. 412.—Pfeif. Mon. Helic. vol. 1, p. 132 (with synonymy).
Ceylon, and Malabar.

A very variable shell, often without bands, too well known to require much illustration.

PLATE CXXXI.
HELIX.

1, 2, 3. H. Feddeni, Blanford, J. Asi. Soc. Beng. 1865, pt. 2 (vol. 30), p. 75 ; Cont. Mal. pt. 5, p. 11.—Pfeif. Mon. Helic. vol. 5, p. 398.
Near Prome, Pegu.

4. H. trifasciata, Chemnitz, Conch. Cab. vol. 11, f. 3018, 3019 : ed. Kuster, Helix, p. 108, pl. 84, f. 20, 21, and pl. 136, f. 13. Pfeif. Mon. Helic. vol 5, p. 76.—H. lævipes, var. Férus., Hist. Moll. pl. 92, f. 4.
Malabar, Tranquebar.

CONCHOLOGIA INDICA.

53

5, 6. **H. Candaharica**, Hutton, J. Asi. Soc. Beng.
1849 (vol. 17, pt. 2), p. 630.—Pfeif. Mon. Helic.
vol. 1, p. 165.—Reeve, Conch. Icon. Helix, f.
456.
Candahar.

7, 10. **H. Shiplayi**, Pfeiffer, Proc. Zool. Soc. 1856,
p. 327 : Mon. Helic. vol. 4, p. 39.
Beypur, Anamullay Hills.

8, 9. **H. asperella**, Pfeiffer, Symb. pt. 3, p. 36 :
Mon. Helic. vol. 1, p. 364: Kust. ed. Chemn. pl.
82, f. 22 to 25.—Reeve, Conch. Icon. Helix, f.
752.
Bundelkhund.

PLATE CXXXII.

HELIX.

3, 4. **H. clathratula**, Pfeiffer, Zeitsch. Malak, 1850,
p. 67 : Mon. Helic. vol. 3. p.115.—Bens. Ann. Nat.
Hist. 1860 (ser. 3, vol. 5), p. 247.—Reeve, Conch.
Icon. Helix, f. 336.—H. puteolus, Bens. Ann. Nat.
Hist. 1853 (ser. 2, vol. 12), p. 92.—Reeve, Conch.
Icon. Helix, f. 1334.
Ceylon.

2, 3. **H. nepos**, Pfeiffer, Proc. Zool. Soc. 1855, p.
91 : Mon. Helic. vol. 4, p. 24.
Ceylon.

5, 6. **H. injussa**, W. and H. Blanford, J. Asi. Soc.
Beng. 1861, vol. 30, p. 355, pl. 1, f. 13 : Cont.
Mal. pt. 2.—Pfeif. Mon. Helic. vol. 5, p. 181.
Coonoor Ghat, Nilgherries.

7. **H. liricincta**, Theobald and Stol. J. Asi. Soc.
Beng. 1871, vol. 40, pt. 2, p. 241, pl. 18, f. 10
(as Conulema).
Near Moulmein.

8, 9. **H. sericata**, Godwin-Austin, Proc. Zool. Soc.
1874.
N. Cachar.
The author, to whose courtesy we are much indebted, proposes to describe this remarkable shell in
the Zoological Proceedings as a Plectopylis.

10. **H. Bensoni**, Von dem Busch in Philip. Abbild.
N. Conch. vol. 1, p. 11, Helix, pl. 1, f. 7.—Pfeif.
Mon. Helic. vol. 1, p. 216.
Bengal (teste Pfeiffer): Khasia Hills.
Will possibly be considered a variety of serrulata.

PLATE CXXXIII.

**MEGALOMASTOMA, RAPHAULUS,
STREPTAULUS, HELICINA, CLOSTOPHIS.**

See previous plate vii, for Megalomastoma, and vi,
for Helicina.

1. **M. funiculatum**, var. Benson.
For typical form see pl. vii. f. 2.

2, 3. **M. pauperculum**, Benson, in Sow. Thesaur.
Conch. vol. 3, pl. 262, f. 22.—Pfeif. Mon. Pneum.
vol. 2, p. 85.
Himalayah.
Figured from the unique type ; very near to the last
species.

4. **R. pachysiphon**, Theobald and Stolic. J. Asi.
Soc. Beng. 1872, vol. xli. pt. 2, p. 329, pl.
11, f. 1.
Near Moulmein, Martaban.

5, 6. **S. Blanfordi**, Benson, Ann. Nat. Hist. ser. 2,
vol. 19 (1857), p. 201.—Pfeif. Mon. Pneum. vol.
2, p. 92.—Sow. Thes. Conch. vol. 3, pl. 265, f. 8, 9.
Near Darjiling in Sikkim Himalayah.

7. **R. chrysalis**, Pfeiffer, Proc. Zool. Soc. 1852, p.
158 (as Cyclostoma): Mon. Pneum. vol. 2, p. 92.—
Bens. Ann. Nat. Hist. ser. 2, vol. 17 (1856), p.
342 (as Aualhus).—Sow. Thes. Conch. vol. 3, pl.
265, f. 6, 7.—Pollicaria c. Gould, Otia, p. 221.
Ava: near Moulmein.

8, 9. **H. scrupulum**, Benson, Ann. Nat. Hist. 1863,
ser. 3, vol. 12, p. 425.—Pfeif. Mon. Pneum. vol.
3, p. 239.
Andaman Islands.
We have figured the unique type.

10. **Clostophis Sankeyi**, Benson, Ann. Nat. Hist.
1860, ser. 3, vol. 5, p. 95.—Pfeif. Mon. Pneum.
vol. 3, p. 12.
The farm-caves near Moulmein.
Only a single and somewhat imperfect specimen
(possibly a monstrosity) has been found : it reminds us
somewhat of an imperfect Opisthostoma.

PLATE CXXXIV.

PTEROCYCLOS (including **SPIRACULUM**, &c.)

See previous plates x, xlix.

1. **P. Feddeni**, Benson. See previous plate v. f. 9.

2, 3, 4. **P. pullatus**, Benson, Ann. Nat. Hist. 1856, ser. 2, vol. 17, p. 227.—Pfeif. Mon. Pneum. vol. 2, p. 31.—Reeve, Conch. Icon. Pter. f. 16.

Akoutong near Irrawadi.

5, 6. **P. (Sp.) Beddomei**, Blanford, J. Asi. Soc. Beng. vol. 38, 1866, pt. 2, p. 31 : Cont. Mal. pt. 6 (as Spiraculum).

Near Vizagapatam, Presidency of North Madras.

7, 10. **P. cotra**, Benson, Ann. Nat. Hist. 1856, ser. 2, vol. 17, p. 228.—Reeve, Conch. Icon. Pter. f. 11.

Moulmein.

8, 9. **P. (Sp.) Avanus**, Blanford, J. Asi. Soc. Beng. 1863, vol. 32, p. 319 : Cont. Mal. pt. 4. (as Spiraculum).

Shan Hills, East of Ava : Kimety Hills.

PLATE CXXXV.

CRASPEDOTROPIS, JERDONIA, LAGOCHEILUS, CYATHOPOMA.

See previous plate lxxxii (for Cyathopoma), and vi (for Lagocheilus).

1, 4. **Cr. cuspidatus**, Benson, Ann. Nat. Hist. 1851, ser. 2, vol. 8, p. 189 (as Cyclostoma).—Pfeif. in Kust. ed. Chemn. Cyclos. pl. 49, f. 21, 22, 23, and Mon. Pneum. vol. 1, p. 313 (as Cycl.), vol. 2, p. 62 (as Cycloph.)—Reeve, Conch. Icon. Cycl. f. 39 (as Cyclophs.)—Cras. c. Blanf. Ann. Nat. 1864, June.

Nilgherries.

2. **Lag. leporinus**, Blanford, J. Asi. Soc. Beng. 1865, pt. 2, vol. 34, p. 82, and Cont. Mal. pt. 5, (as Cycloph. section Lagoe.).

Akoutong, Pegu.

3. **J. ? Phayrei**, Theobald, J. Asi. Soc. Beng. 1870, vol. 39, pt. 2, p. 396.

Shan : South Canara.

When in very fine condition the hairs are much elongated.

5, 6. **J. trochlea**, Benson, Ann. Nat. Hist. 1851, ser. 2, vol. 8, p. 189 (as Cyclos.).—Cyclostomus tr. Pfeif. Mon. Pneum. vol. 2, p. 116 : Kust. ed. Chemn. Cyclos. pl. 49, f. 29, 30.

Nilgherries.

7. **Cy. procerum**, Blanford, Journ. Conch. 1868, p. 262, pl. 12, f. 8.

Beypoor, Malabar.

8, 9. **Cy. Kolamullionse**, W. and H. Blanford, J. Asi. Soc. Beng. 1861, vol. 30, p. 351, pl. 1, f. 4, and Cont. Mal. pt. 2 (as ? Jerdonia K.) : Ann. Nat. 1861, June.—Pfeif. Mon. Pneum. vol. 3, p. 28 (as Cyclotus).

Kolamullay Hills.

Mr. W. Blanford has kindly identified our specimen, which does not quite agree with the figure in the Journal de Conchyliologie (1868, pl. 12, f. 5).

10. **Cy. Coonoorense**, Blanford, Journ. Conch. 1868, p. 261, pl. 12, f. 6.

Krore Mund, top of Nilgherries.

PLATE CXXXVI.

MYCHOPOMA and DITROPIS.

1, 4. **M. hirsutum**, Beddome, in Blanford's Cont. Mal. Ind. pt. 10, in Jour. Asi. Soc. Beng. 1869, vol. 38, pt. 2, p. 132, pl. 16, f. 5, as Cyclophorus (Myc.).

Calcad and Myhendra Hills.

2, 3. **M. limbiferum**, Blanford, Journ. Asi. Soc. Beng. 1869, vol. 38, pt. 2, p. 133, pl. 16, f. 4 : Cont. Mal. pt. 10 as Cyclophorus (Myc.).

Tops of Pulney Hills.

5, 6. **D. planorbis**, Blanford, Journ. Asi. Soc. Beng. 1869, vol. 38, pt. 2, p. 126, pl. 16, f. 1, as Cyclophorus (Dit.).

Calcad Hills, limits of Travancore.

7, 10. **D. convexus**, Blanford, Journ. Asi. Soc. Beng. 1869, vol. 38, pt. 2, p. 128, pl. 16, f. 3 : and Cont. Mal. pt. 10, and as Cyclophorus.

Calcad Hills, limits of Travancore.

8, 9. **D. Beddomei**, Blanford, Journ. Asi. Soc. Beng. 1869, vol. 38, pt. 2, p. 127, pl. 16, f. 2, and Cont. Mal. pt. 10, as Cyclophorus (Dit.).

Travancore.

PLATE CXXXVII.

NAVICELLA.

1, 4. **N. compressa**, Pearson's MSS. in Benson, J. Asi. Soc. Beng. 1836, vol. 5, p. 749.—N. lineata, var. Sow. Thes. Conch. vol. 2, pl. 118, f. 25.

River Hoogly.

The Patella Aponogetonis of Vahl (Skrift. Nat. Selskab. vol. 4, pt. 2, p. 153) is probably the young of this, or of the next species.

2, 3, 7. **N. cœrulescens**,' Récluz, in Sow. Thes.
Conch. vol. 2, p. 550, pl. 118, f. 29.—Reeve,
Conch. Icon. Navic. f. 29, 37.—N. orientalis, Reeve,
do. f. 33 (young).—N. tessellata, Bens. (not well of
Lam.) Journ. Asi. Beng. Soc. 1836 (vol. 5),
Ganges, Bengal.
This probably was the Gangetic form of the La-
marckian elliptica referred to by Troschel (Wiegm.
Arch. Nat. 1837).

5, 6. **N. reticulata**, Reeve, Conch. Icon. Navic.
f. 20; and N. eximia, f. 26.
Ceylon.

8, 9. **N. Livesayi**, Dohrn, Proc. Zool. Soc. 1858,
p. 35.
Ceylon.
Drawn from the originals in the late Cumingian
collection.

10. **N. cœrulescens**, var., Récluz, in Sow. Th. Conch.
vol. 2, pl. 118, f. 36, 38.
Hindostan.

PLATE CXXXVIII.

CORBICULA.

1, 4. **C. Bensoni**, Deshayes, Proc. Zool. Soc. 1854,
p. 345: Cat. Brit. Mus. Vener. p. 225.
River Jumna.
Apparently rare, and notable for its general smoothness; it has occasionally indistinct, interrupted
radiating lines.

2, 3. **C. Cashmirensis**, Deshayes, Proc. Zool. Soc.
1854, p. 344: Cat. Brit. Mus. Vener. p. 224.
Beloochistan, Avantipura, Cashmire, &c. &c.

5, 6. **C. regularis**, Prime, Proc. Zool. Soc. 1860,
p. 321.
Madras, Deccan, in brackish water.

7, 10. **C. striatella**, Desh. Proc. Zool. Soc. 1854,
p. 344: Cat. Brit. Mus. Vener. p. 224.—Hanley,
Photog. Conch.—Prime, Ann. Lyc. 1864, vol. 8,
p. 74, f. 22.—C. violacea, Prime, do. 1861, p. 28,
teste Prime.
Pondicherry, &c.

8, 9. **C. occidens**, "Benson" in Desh. Cat. Brit.
Mus. Vener. p. 225.—Hanley, Photog. Conch.—
Prime, Ann. Lyc. N.Y. 1866, vol. 8, p. 220, f. 51.
Sikkim : Bundelkhund, &c.
Has sometimes (yet rarely) linear rays of rufous
brown on its yellowish ground colour.

PLATE CXXXIX.

DIPLOMMATINA.

See previous plates cxix, cxx.

1. **D. tumida**, Godwin-Austen, J. Asi. Soc. Beng.
1870, p. 6, pl. 2, f. 2.
Khasi Hills.

2, 3. **D. scalaris**, Blanford, J. Asi. Soc. Beng. 1865,
p. 79, pl. 2, f. 2 : Cont. Mal. pt. 9.
Garo Hills, W. of Khasi.

4. **D. parvula**, Godwin-Austen, J. Asi. Soc. Beng.
1870, p. 5, pl. 1, f. 5.
Khasi Hills.

5, 6. **D. Huttoni**, Pfeiffer, Proc. Zool. Soc. 1852, p.
157 : Mon. Pneum. vol. 1, p. 123 : Kust. ed. Chemn.
Cyclos. pl. 48, f. 36, 37.
Mussoorie.

7. **D. Jatingana**, Godwin-Austen, J. Asi. Soc. Beng.
1870, p. 1, pl. 1, f. 1.
Khasi Hills.

8, 9. **D. Puppensis**, Blanford, J. Asi. Soc. Beng.
1862, vol. 32, p. 324 ; Ann. Nat. Hist. 1864, June.
Puppa Hill, Upper Birmah.

10. **D. insignis**, Godwin-Austen, J. Asi. Soc. Beng.
1870, p. 6, pl. 2, f. 1.
Shan States : Khasi Hills.

PLATE CXL.

DIPLOMMATINA.

1. **D. nana**, Blanford, J. Asi. Soc. Beng. 1865, pt. 2,
vol. 34, p. 85.
Akoutong, Thondoung, Yenandoung, Henzada
district, Pegu : near Moulmein.

2, 3. **D. diplocheilos**, Benson, Ann. Nat. Hist. ser. 2,
vol. 19 (1857) p. 202.—Pfeif. Mon. Pneum. vol. 2,
p. 16, Novit. Conch. vol. 1, pl. 37, f. 13, 14, 15.
Teria Ghat, Khasi Hills.

4. **D. carneola**, Stoliczka, J. Asi. Soc. Beng. 1871,
vol. 40, pt. 2, p. 152, pl. 6, f. 3.
Damotha, near Moulmein.

5, 6. **D. pachycheila**, Benson, Ann. Nat. Hist. 1857,
ser. 2, vol. 19, p. 203.—Pfeif. Mon. Pneum. vol. 2,
p. 16 : Novit. Conch. vol. 1, pl. 37, f. 16, 17, 18.
Darjiling.

7. **D. angulata**, Theobald and Stolic. J. Asi. Soc.
Beng. 1872, vol. 41, pt. 2, p. 331, pl. 11, f. 3.
Near Moulmein, Martaban.

8, 9. **D. folliculus**, Pfeiffer, Mon. Pneum. vol. 1,
p. 122 (previously in Symbol. Helic. pt. 3, p. 83,
as Bulimus), vol. 2, p. 10 ; Kust. ed. Chemn. Cyclos.
pl. 48, f. 32, 33.
Landour ; Simla.

10. **D. polyplouris**, Benson, Ann. Nat. Hist. ser. 2,
vol. 19, 1857, p. 203.—Pfeif. Mon. Pneum. vol. 2,
p. 11.—G. Austen, J. Asi. Soc. Beng. 1870, p. 4,
pl. 1, f. 4, var. lævior.
Nanclai ; Ponji ; Sandowny.

PLATE CXLI.
DIPLOMMATINA.

1. **D. Kingiana**, W. and H. Blanford, J. Asi. Soc.
Beng. 1861, vol. 30, p. 348, pl. 1, f. 2.—Arinia
Kin. Pf. Mon. Pneum. vol. 3, p. 91.—D. (Nicida)
King. Blanf. J. Conc. 1868.
Kohamulhy Hills, near Trichinopoli, Southern
India.

2. **D. liricincta**, Blanford, Journ. Conch. 1868,
pl. 14, f. 5, as D. (Nicida) lir.
Khandullah, between Bombay and Poona, in
Syhadri Hills.

3. **D. Pulncyana**, Blanford, Journ. Conch. 1868,
pl. 14, f. 2, as D. (Nicida) Pul.
Pulney Hills, Southern India.

4. **D. Nilgirica**, W. and H. Blanford, J. Asi. Soc.
Beng. 1860, vol. 29, p. 124, and 1861, p. 348,
pl. 1, f. 1.—Arinia Nil. Pf. Mon. Pneum. vol. 3,
p. 91.—D. (Nicida) Nil. Bl. Journ. Conch. 1868,
pl. 14, f. 1.
Near Pykara, Nilgherries.

5. **D. nitidula**, Blanford, Journ. Conch. 1868,
pl. 14, f. 8, as D. (Nicida) nit.
Kulputty Hills, Wynaad, Nilgherries.

6. **D. crispata**, Stoliczka, J. Asi. Soc. Beng. 1871,
vol. 40, pt. 2, p. 153, pl. 6, f. 4, as D. (Palaina) C.
Damotha, near Moulmein.

7, 8. **D. Richthofeni**, Theobald & Stol. J. Asi.
Soc. Beng. 1872, vol. 41, pt. 2, p. 331, pl. 11, f. 4.
Near Moulmein, Martaban.

9. **D. Fairbanki**, Blanford, Journ. Conch. 1868,
pl. 14, f. 4, as D. (Nicida) F.
The unique example being lost, we have copied our
figure from the work referred to.

10. **D. scalaroides**, Theobald, J. Asi. Soc. Beng.
1870, vol. 39, pt. 2, p. 399, pl. 18, f. 5.
Mandalay, Birmah.

PLATE CXLII.
LEPTOPOMA, PTEROCYCLOS.

See previous plates vi, cv (for Leptopoma), and v, xlix,
cxxxiv (for Pterocyclos).

1. **L. apicatum**, Benson, Ann. Nat. Hist. 1856 (ser.
2, vol. 18), p. 95.—Pfeif. Mon. Pneum. vol. 2,
p. 73 (copied from Bens.).—Reeve, Conch. Icon.
Lept. f. 33.
Ceylon.

2. **L. olatum**, Pfeiffer, Proc. Zool. Soc. 1852, p.
159, and in Kust. ed. Chemn. Cyclost. pl. 32, f. 16,
17 (as Cyclost.): Mon. Pneum. vol. 1, p. 117 (as
L.).—Reeve, Conch. Icon. Lept. f. 3.
Ceylon.

3. **L. flammoum**, Pfeiffer, Proc. Zool. Soc. 1834,
p. 127 (as Cyclost.): Mon. Pneum. vol. 2, p. 76
(as Lept.).—Reeve, Conch. Icon. Lept. f. 47 a.
Ceylon.
Figure 47 b of Reeve is not satisfactory. We speak
from the type. Our own figure is taken from a banded
variety.

4. **L. orophilum**, Benson, Ann. Nat. Hist. 1853,
ser. 2, vol. 11, p. 106 (as Cyclost.): Mon. Pneum.
vol. 2, p. 77 (as L.).—Reeve, Conch. Icon. Lept.
f. 51. L. pœcilum, Pfeiffer, Proc. Zool. Soc. 1854,
p. 302 (as Cyclost.): Mon. Pneum. vol. 2, p. 76.—
Reeve, Conch. Icon. Lept. f. 46.
Ceylon.
The name orophilum has priority ; pœcilum, however, is the more clearly defined species.

5, 6. **P. ater**, Stoliczka, J. Asi. Soc. Beng. 1872,
vol. 40, pt. 2, p. 149, pl. 6, f. 2.
Kuengan, near Moulmein.

8, 9. **P. bifrons**, Pfeiffer, Proc. Zool. Soc. 1855,
p. 117 : Mon. Pneum. vol. 2, p. 30.—Reeve,
Conch. Icon. Pter. f. 1.
Ceylon.

7, 10. **P. parvus**, Pearson. See plate 5, f. 3 for var.
Assamensis.

Our figure 7 represents the smaller and more
typical form of parvus, 10 that of the variety which
Benson mistook for the P. Albersi. The provision-
ally named but undescribed P. Arakanensis (Blanford
in J. Asi. Soc. Beng. 1865, p. 98 : Cont. Mal. Ind.
pt. 5) differs little from the variety Assamensis.

PLATE CXLIII.

CYCLOPHORUS.

See other plates i to iv, xxxiii, xxxiv, xlvii, xlviii, civ, cv, cliv, clv.

1, 4. **C. annulatus**, Troschel, in Pfeif. Zeits. Malak. 1847, p. 150.—Pf. Mon. Pneum. vol. 1, p. 98; Kust. ed. Chemn. Cyclost. pl. 29, f. 14, 15.

Koondah Mountains; Ceylon.

In the earlier description, which differs considerably from that in Pfeiffer's Monographs, an interrupted peripheral band (as in Kust. pl. 22) is suggested; we prefer, then, to assign the name annulatus to that much more ringed form which we have figured.

2, 3. **C. parma**, Benson, Ann. Nat. Hist. ser. 2, vol. 18 (1856), p. 94.—Pfeif. Mon. Pneum. vol. 2, p. 55.

Ceylon.

Allied to cratera and cytopoma; the very closely coiled operculum of the former has (it is said) two more whorls; the peristome of the latter is not double.

5, 6. **C. tristis**, Blanford, J. Asi. Soc. Beng. vol. 38, p. 134, pl. 16, f. 9 (as Pterocyel.).

S. Canara.

The discovery of the operculum forces us to remove this abnormal species from that genus to which it was first assigned.

7, 10. **C. Shiplayi**, Proc. Zool. Soc. 1856, p. 337; Mon. Pneum. vol. 2, p. 68.—Reeve, Conch. Icon. Cyclop. f. 85.

Nilgherries.

Figured from the original types in the British Museum, which are very possibly immature.

8, 9. **C. Inglisianus**, Stoliczka, Journ. Asi. Soc. Beng. 1871, vol. 40, pt. 2, p. 148, pl. 6, f. 1.

Damotha, near Moulmein.

PLATE CXLIV.

CYCLOPHORUS.

1. **C. fulguratus**, var. Pfeiffer. See previous figure on plate 1.—C. fulguratus, Reeve, Conch. Icon. Cyclop. f. 35, c. d.—Pfeif. Novit. pl. 98, f. 1, 2.

The original type of fulguratus was the young shell figured by Pfeiffer in his monograph in Kuster's edition of Martini and Chemnitz (Cycl. pl. 15, f. 9, 10), and by Reeve in his Iconica (Cyclop. f. 35, a, b).

2. **C. Theobaldianus**, var. Benson. Birmah.

Almost a link between Theobaldianus and pracosus.

3, 4. **C. Phayrei**, Theobald, MSS. Moulmein, Birmah.

Reminds one of Ceylanicus, and a little of Haughtoni, of which one writer considers it a variety.

5. **C. alabastrinus**, Pfeiffer, Proc. Zool. Soc. 1854, p. 126 (as Cyclost.); Mon. Pneum. vol. 3, p. 41; Novit. Conch. vol. 1, pl. 1, f. 4, 5.

"Ceylon"?

We doubt both the locality and the distinctiveness of this dead shell, but figure the better of the specimens in the British Museum.

6. **C. ophis**, Hanley, Pr. Zool. Soc. 1875. Tenasserim.

Somewhat allied to C. tuba, but quite distinct.

7. **C. serratizona**, Thorp, MSS. Upper Salwen (Theobald).

The jagged edge of the white band forms a conspicuous, yet perhaps not permanent feature. The shell, which has a white aperture, and a large umbilical area, comes between Phayrei and polynema; the faint close spiral ridge seem confined to the upper disc. Except in shape it might be taken for C. labiosus.

PLATE CXLV.

CYCLOPHORUS, ALYCÆUS, OMPHALOTROPIS, CATAULUS, CYATHOPOMA.

See for Alycæus plates xci to xcvii, ciii; for Cataulus cvi, cxlvi; for Cyathopoma lxxxii, cxxxv.

1, 4. **Al. expatriatus**, Blanford, J. Asi. Soc. Beng. 1860, vol. 29, p. 123.—Pfeif. Mon. Pneum. vol. 3, p. 52.

Nebbowuttum Ghat, north of Nilgherries, and var. from Shevroys.

2, 3. **Al. Kurzianus**, Theobald and Stoliczka, J. Asi. Soc. Beng. 1872, vol. 41, pt. 2, p. 330, pl. 11, f. 3.

Nattoung, Prome.

5, 7. **Cyc. subplicatulus**, Beddome, Pr. Zool. Soc. 1875, p. 452, pl. 53, f. 26, 27.

Ceylon (teste Beddome).

6. **Cat. marginatus**, Pfeiffer, Proc. Zool. Soc. 1853,

p. 52; Mon. Pneum. vol. 2, p. 90.—Sow. Thes.
Conch. vol. 3, pl. 264, f. 4, 5.
Ceylon.

8. **Cyat. Ceylanicum**, Beddome. Proc. Zool. Soc.
1875, p. 450, pl. 52, f. 20.
Near Rambodda Falls, Ceylon.

9. **Cyat. vitreum**, Beddome, Proc. Zool. Soc.
1875, p. 449, pl. 53, f. 21, 22.
Tinnevelly district, S. India.

10. **O. disterminra**, Benson, An. Nat. H. 1865, Dec.
—Pfeif. Mon. Pn. vol. 3, p. 178.

Andamans (one specimen, here figured).

PLATE CXLVI.

CATAULUS, CREMNOCONCHUS.

See previous plate (for Cataulus) cvi.

1. **Cat. leucocheilus**, Adams and Reeve, in Sow.
Thes. C. vol. 3, pl. 264, f. 14.
Ceylon (coll. Beddome).

2. **Cat. recurvatus**, Pfeiffer, Proc. Zool. Soc. 1862,
p. 116, pl. 12, f. 2; Mon. Pneum. vol. 3, p. 88.—
Sow. Thes. Conch. vol. 3, pl. 264, f. 16.
Anamallay forest, foot of Nilgherries.

3. **Cat. eurytrema**, Pfeiffer, Proc. Zool. Soc. 1852,
p. 145, pl. 13, f. 5 (badly).—Sow. Thes. Conch.
vol. 3, pl. 264, f. 17.
Ceylon; Travancore (Beddome).

4. **Cat. Nietneri**, Nevill, J. Asi. Soc. Beng. vol.
39, pt. 2 (1871), pl. 1, f. 7, 7, a.
Ceylon.

The original was drawn from a depauperated
variety.

5. **Cat. pyramidatus**, Pfeiffer, Proc. Zool. Soc.
1852, pl. 15, f. 4 (badly); Mon. Pneum. vol. 2,
p. 88. Sow. Thes. Conch. vol. 3, pl. 264, f. 10.
Ceylon.

The distinctness of this from Austenianus may be
doubted.

6. **Cr. Syhadrensis**, Blanford, An. Nat. Hist.
ser. 3, vol. 12, p. 184.
On hills opposite Bombay.

7. **Cr. Fairbanki**, Blanford.

An accident to our manuscript at the time of going

to press prevents our saying where (if at all) this
species has been published.

8, 9. **Cr. conicus**, Blanford, J. Asi. Soc. Beng.
1870, vol. 39, pt. 2, p. 10, pl. 3, f. 3, 4.
Torna, near Poona.

10. **Cr. carinatus**, Layard, Proc. Zool. Soc. 1854,
p. 94 (as Auculotus).—Blanf. J. Asi. Soc. Beng.
1870, vol. 39, pt. 2, p. 10, pl. 3, f. 3, 4.
Mahableshwar Hills, Bombay Presidency.

Originally described from a young shell.

PLATE CXLVII.

SOPHINA, HYPSELOSTOMA.

See plate viii for Hypselostoma.

1, 4. **S. forabilis**, Benson, Ann. Nat. Hist. 1
(as Helix), amended 1860 (ser. 3, vol. 5), p. 27.
—Pfeif. Mon. Helic. vol. 5, p. 112 (as H.).—
Stolic. J. Asi. Soc. Beng. 1871 (vol. 40, pt. 2),
p. 257, pl. 19, f. 10.—Pfeif. Mon. Hel. vol. 7,
p. 117 (as Hel.).
Phai Than, Tenasserim Valley, and Damatha
Cavern, near Moulmein.

2, 3. **S. Calias**, Benson, An. Nat. Hist. 1859 (as
Helix), ser. 3, vol. 5, p. 473, amended 1860
(ser. 3, vol. 5), p. 26.—Pfeif. Mon. Helic.
vol. 5, p. 112 (as H.).
Farm caves, near Moulmein.

5, 6. **S. schistostelis**, Benson, Ann. Nat. Hist.
ser. 3, vol. 3, 1859, p. 473 (as Helix), amended
An. Nat. 1860, ser. 3, vol. 5, p. 27 (as S.).—
Pfeif. Mon. Helic. vol. 5, p. 111 (as H.).—S.
Calias, Stolic. J. Asi. Soc. Beng. 1871, vol. 40,
pt. 2, p. 255, pl. 19, f. 1, 4, 7, 9, text copied
Pfeif. Mon. Hel. vol. 7, p. 116.
Near Moulmein.

7. **S. Calias**, Benson, var. discoidalis.—S. discoidalis, Stolic. J. Asi. Soc. Beng. 1871, vol. 40,
pt. 2, p. 258, pl. 19, f. 15, 11, 12.—Pfeif. Mon.
Hel. vol. 7, p. 117 (as Helix).
On limestone hills, S. of Moulmein.

Benson's type of the true Calias was quite different
from that supposed to be it by Stoliczka; hence his
erroneous introduction of S. discoidalis to nomen-
clature.

8, 9. **S. conjungens**, Stoliczka, J. Asi. Soc.
Beng. 1871, vol. 40, pt. 2, p. 259, pl. 19, f. 6, 13,
text copied Pfeif. Mon. Hel. vol. 7, p. 118.
South of Moulmein.

10. **H. Dayanum**, Stoliczka. J. Asi. Soc. Beng.
1871, vol. 40, pt. 2, p. 172, pl. 7, f. 2.
Damotha, near Moulmein.

Our figure is merely copied from the one here
cited.

PLATE CXLVIII.

BULIMUS.

See previous plates xix to xxiii, lxxix, lxxx.

1, 4. **B. (Hapalus) Munipurensis**, Godwin-
Austen, Proc. Zool. Soc. 1872, p. 516, pl. 30,
f. 8.
Hengdan Peak in the Munipur boundary.

2, 3. **B. Caleadonsis**, Beddome MSS. in Blanf.
J. Asi. Soc. Beng. 1870, vol. 39, pt. 2, p. 18.
Travancore.

5. **B. Mavortius**, Reeve, Conch. Icon. Bulim.
f. 561.—Pfeif. Mon. Pneum. vol. 3, p. 423.
Ceylon.

6. **B. Bengalensis**, Lamarck. Anim. s. Vert. See
pl. 80, f. 7.
We here figure the typical two-banded form delineated in Delessert's folio.

7. **B. (Hapalus) Khasianus**, Godwin-Austen,
Proc. Zool. Soc. 1872, p. 516, pl. 30, f. 7.
Khasi, Jaintea and Naga Hills.

8. **B. adumbratus**, Pfeiffer, Proc. Zool. Soc. 1854,
p. 291: Mon. Pneum. vol. 4, p. 472.
Ceylon.

Our praetermissus, var. (pl. 19, f. 4) runs into
this: the type here delineated is in the British
Museum.

9. **B. Ceylanicus**, var.

A very beautifully painted form, which will by
some be considered a distinct species, by others referred to physalis.

10. **B. Andamanicus**, Thorp, MSS.
Andaman Islands.

This manuscript name has been tardily accepted,
for although the shell has been regarded by some

as the contrarius of Müller, his description does not
at all apply to our specimen. The aperture is white
with a broad purple lake band above the pillar. For
the sinistral Bulimi of Asia the student is referred to
Von Martens in the zoology of "Die Preussiche Ex-
pedition nach Ost-Asien."

PLATE CXLIX.

HELIX.

See previous plates xiii to xvi, xxv to xxxii, l to lxiv.
lxxxiii to xc, cxi, cxii, cxxvii to cxxxii.

1. **H. hobesceens**, Blanford, J. Asi. Soc. Beng.
1865 (vol. 36), p. 34, and Cont. Mal. pt. 6, p. 1
(as Nanina).—Pfeif. Mon. Helic. vol. 5, p. 78.
Anamallay Hills, S. India.

2, 3. **H. subcornca**, Pfeif. Proc. Zool. Soc. 1861,
p. 29: Mon. Helic. vol. 5, p. 105: Mal. Blat.
1859, p. 232.
Phie Than (Theobald).

Very near H. resplendens of Philippi, and probably
more abundant in Siam than at Phie Than.

4. **H. Phidias**, Thorp, MSS.
Upper Ouvah, Ceylon (F. Layard).

Pfeiffer had confused this and hyphasma in the
Cumingian collection: its smooth marble-like surface
does not agree with the expression "sulcis rotundi-
oribus spiralibus quasi texta."

5, 6. **H. lixa**, Blanford, J. Asi. Soc. Beng. 1865,
vol. 36, p. 35 (as Nanina).—Pfeif. Mon. Helic.
vol. 5, p. 78.
Anamallay Hills, Southern India.

7. **H. Travancorica**. See previous plate 30, f.
5, 6.

8, 9. **H. (Plectopylis) Shanensis**, Stoliczka, J.
Asi. Soc. Beng. 1873, vol. 42, p. 170 (as Plect.).
Shan States.

Not unlike Kuster's figure of refuga. The three
labial plicae (the middle being remote), and the two
labial notches are the salient external characters.
We do not find it referred to by Godwin-Austen in
his valuable paper in the Zoological Proceedings for
1874.

10. **H. Footei**, Stoliczka, J. Asi. Soc. Beng. 1873,
vol. 42, p. 170 (as Trochia).
Poona and Belgaum.

PLATE CL.

HELIX.

1, 2. H. (Plect.) Beddomeæ, Hanley.
Ceylon or Southern India.

The only specimen known was collected by Col.
Beddome, in honour of whose amiable wife the species
has been named.

3. H. corylus, Reeve, Conch. Icon. Hel. f. 1439.—
Pfeif. Mon. Hel. vol. 4, p. 55.
Ceylon.

Very close to partita, but darker and less sculptured. Our drawing is from the type now in the
National Museum.

4. H. prospera, Albers, Mal. Blät. 1857 (vol. 4),
p. 13, pl. 1, f. 7, 8.—Pfeif. Mon. Helic. vol. 4,
p. 197, and vol. 5, p. 267.
Ceylon.

Our figure is a mere copy from the published en-
graving : we have never seen the species.

5, 6. N. liratula, Pfeiffer, Proc. Zool. 1860, p. 35 :
Mal. Blät. vol. 7, p. 234 : Mon. Hel. vol. 5, p. 182.
Ceylon, 6,000 feet above the sea-level.

The original types are here delineated.

7. H. immerita, Blanford, J. Asi. Soc. Beng.
1870 (vol. 39, pt. 2), p. 17 (as Nanina, section
Ariophanta).—Pfeif. Mon. Helic. vol. 7, p. 128.
South Canara.

The original type is here figured.

8. H. novella, Pfeiffer, Proc. Zool. Soc. 1851,
p. 50 : Mon. Hel. p. 34.—Reeve, Conch. Icon.
Hel. f. 1294.
Ceylon.

The drawing is taken from the original specimens.

9. H. verrucula, Pfeiffer, Proc. Zool. Soc. 1851,
p. 56 : Mon. Helic. vol. 4, p. 40.—Reeve, Conch.
Icon. Hel. f. 1327.
Ceylon.

Cuming's type, now in the British Museum, is here
represented.

10. H. daghoba, W. and H. Blanford, J. Asi. Soc.
Beng. 1861 (vol. 30), p. 356, pl. 2, f. 2.—Pfeif.
Mon. Helic. vol. 5, p. 219.
Patchamullies and Kalryenmullies.

Our figure is taken from Blanford's lithograph,
aided by a poor specimen of our own. The shell
reminds one of a very broad bidenticulata.

PLATE CLI.

PLANORBIS, AMNICOLA, BITHINIA.

See previous plates xxxix, xl, xcix, for Planorbis, and
xxxvii, xxxviii, for Bithinia.

1, 2, 3. P. elegantulus, Dohrn, Proc. Zool. Soc.
1858.
Ceylon.

4, 7. P. Stelznori, Dohrn, Proc. Zool. Soc. 1858.
Ceylon.

5, 6. P. Merguiensis, Philippi.
Mergui, Birmah.

Our specimens were received from Philippi thus
named, but we cannot find where he has described
them.

8, 9. A. parvula, Hutton, J. Asi. Soc. Beng. 1849,
(vol. 17, pt. 2) and vol. 18, pt. 2, p. 655 (as
Paludina).—Bithinia globula, Lea, Proc. Philad.
1856, vol. 8, p. 110, and Journ. Ac. Philad. new
ser. vol. 6, p. 119, pl. 22, f. 12 (=Olus. Un.
vol. 11) probably.
Chunamun, Afighanistan ; Ahmednuggur.

We are indebted to Major Hutton for our specimens, which are undistinguishable from some Sicilian
shells which Mr. Hanley has received from Benoit,
as his vestita, a name not to be found in Kobelt's
catalogue of European fluviatiles.

10. B. costigera, Beck in Kuster's ed. Chemn.
Conch. Palud. p. 35, pl. 7, f. 18, 19.
Valvata sulcata, Eyd. and Soul. Voy. Bonite Zool.
p. 517, pl. 31, f. 19, 20, 21.
Bengal ; Pondicherry ; Ceylon.

We believe the Cyclostoma gradatum of Pfeiffer
(Zool. Proc. 1854, p. 303) should be referred to this
shell, but cannot find the types. The Turbo marginatus of Chemnitz may, also, be identical, but is too
ill-defined to merit precedence.

PLATE CLII.

VITRINA.

See previous plates lxv, lxvi.

1, 4. **V. monticola**, Benson, MSS. in Pfeif. Mon. Helic. vol. 2, p. 497; Kust. ed. Chemn. Vit. pl. 2, f. 6, 7, 8.—Reeve, Conch. Icon. Vit. f. 11.
Landour, Himalayah.

The surface is highly polished.

2, 3. **V. cassida**, Hutton and Benson, J. Asi. Soc. Beng. vol. 7 (1838), p. 214.—Pfeif. Mon. Helic. vol. 3, p. 2.—Reeve, C. Icon. Vitr. f. 10 (from Benson's specimen).
Himalayah.

The surface is dull, and shows in the type here figured, not merely concentric folds, but, also, some faint and distant spiral striæ.

5. **V. venusta**, Theobald, J. Asi. Soc. Beng. 1870, vol. 39, pl. 2, p. 400.
Arrcan Hills, between Tongoop and Prome.

The aspect of the under side reminds us of a miniature gigas.

6. **V. solida**, Godwin-Austen, Proc. Zool. Soc. 1872, p. 518, pl. 30, f. 10 (as Helicarion).
Hengdan Peak, N. Cachar Hills.

7. **V. Birmanica**, Philippi, Zeits. Malak. 1847, p. 65.—Pfeif. Mon. Hel. vol. 2, p. 498.—Reeve, Conch. Icon. Vit. f. 59.
Mergui, Birmah.

Our drawing is from the same shell figured by Reeve from the collection of Mr. Cuming, who, I think, had it from the author. Yet the term " depressa " is scarcely suitable for the specimen.

8, 9. **V. heteroconcha**, H. Blanford, J. Asi. Soc. Beng. 1871, vol. 40, pt. 2, p. 45, pl. 2, f. 8 (as Helicarion).—Pfeif. Mon. Hel. vol. 7, p. 9.
Darjiling.

10. **V. membranacea**, Benson, An. Nat. H. 1859, ser. 2, vol. 12, p. 93. Pfeif. Mon. Helic. vol. 3, p. 792.—Reeve, Conch. Icon. Vit. f. 78.
Baleadura, Ceylon.

Our figure is taken from the shell delineated by Reeve, which has not arrived at the full size stated by Benson, but is the only specimen attainable.

PLATE CLIII.

MELANIA.

See previous plates lxxi to lxxv, ex.

1. **M. Reevei**, Brot, Cat. Mélan. 1862, p. 46, name for M. baltcata, Reeve (not Phil.) Conch. Icon. Melan. f. 144, b (scarcely 144, a).
Birmah.

The type in the British Museum is an immature form of the shell we had intended to call Goliah (pl. 72, f. 3), and the young may prove the too briefly described M. humerosa of Gould, from Mauko, Tavoy (Proc. Bost. Nat. H. 1847, vol. 2, p. 219).

2. **M. præmordica**, Tryon, American J. Conch. vol. 2, pt. 2, 1866, p. 111, pl. 10, f. 3.
Birmah, Pegu.

Apparently very rare; our sole specimen was a broken one; so the outer lip has been corrected in accordance with Tryon's outline.

3. **M. pagodula**, Gould, Proc. Bost. J. Nat H. 1847, vol. 2, p. 219; Otia Conch. p. 200.—Reeve, Conch. Icon. lo, f. 10 (as lo).
Thoungyin River (branch of Salwen), Birmah.

Our figure is taken from the supposed type in the British Museum, purchased from H. Cuming.

4. **M. Reevei**, var. indeicata.

5, 6, 7. All from the Shan province. Figure 5 is the M. variabilis, var. turrita of Theobald in the J. Asi. Soc. Beng. 1865 (vol. 34, pt. 2, pl. 9, f. 6; f. 7 his variabilis, var. vittata (do. f. 4); f. 5 a form somewhat allied to the first, but with many more rows of granules. They may possibly all be forms of læveata, or possibly all distinct species, but pending the monograph of Melania by Dr. Brot we will not venture to pronounce an opinion.

PLATE CLIV.

UNIO.

See previous plates x, xi, xii, xli to xlvi, cvii.

1. **U. parma**, Benson in Conch. Icon. Unio, f. 514.
Irawadi and Tenasserim River, Pegu.

2. **U. parma?** var. Benson.
   Bhamao.

   May possibly be a distinct species, but having seen but one specimen we do not venture to decide.

3. **U. rugosus**, Gmel. Syst. Natur. f. 3222. for Chemn. Conch. Cab. vol. 10, f. 1649 (= Kust. ed. Chemn. Un. pl. 97, f. 5).
   Coromandel.

   Between rancileuta and scobina. A single valve has been lately obtained which more precisely resembles the figure of Chemnitz.

4. **U. Mandelayanus**, Theobald, J. Asi. Soc. Beng. 1873, p. 208, pl. 17, f. 2.
   Mandelay, Birmah.

   Specimens are occasionally of a rich grass-green.

5. **U. macilentus**, var.
   Surat; near Chinsoor; Peua Gunga.

6. **U. Tavoyensis**, Gould, Proc. Bost. N. H. vol. 1, p. 140; Otia Con. p. 131. —Reeve, Conch. Icon. Unio, f. 49.
   Tavoy.

   Our specimen was sent by Gould to Benson.

7. **U. Tavoyensis?** var.—U. Tavoyensis, Kust. ed. Chem. Unio, pl. 48, f. 2.
   Birmah.

   Very much larger and rounder than the typical form.

   India is the reputed habitat of many Uniones which modern researches have failed to discover there. Amongst them may be specified Spengler's crassus (from Tangiers), his comus (=tumidus, from Europe), his radiatus (from North America), his nodosus (=the European pictorum), &c. Kydous in Guérin's Magasin de Zoologie (1838, pl. 118) indicates U. Kerandreni (f. 1) and U. Gaudichandi (f. 3), as from Bengal (the first may, perchance, be U. leionos, but looks more like a Cape shell), and U. Gerbidoni (f. 2) from Coromandel; these two last remind us of U. Nilotieus and U. lithophagus, both from the Nile. The supposed Indian U. digitiformis of Sowerby comes from China; U. Bengalensis of Lea from the Philippines. U. velaris of Benson and U. delphinus of Spengler must, also, be expunged from our list.

PLATE CLV.

UNIO, CORBICULA, CYCLAS, PISIDIUM, TRICULA, ACHATINA.

See previous plates cxxxviii for Corbicula, and xvii, xviii, xxxv, xxxvi, lxxviii, cii, for Achatina.

1. **T. montana**, Benson, Calcutta J. Nat. H. 1842; Am. Nat. H. 1862, Dec.
   From a stream at the head of Bheentál, on stems of a water iris.

2. **U. Bhamaoensis**, Theob. J. Asi. Soc. Beng. 1873, pt. 2, p. 207, pl. 17, f. 1.
   Near Bhamao, and from Western Promo, Pegu.

3. **U. Vulcanus**, Hanley, Proc. Zool. Soc. 1875.
   Birmah or Pegu.

   The English editor has lost the precise locality of this lovely species.

4. **A. senator**, Hanley, Proc. Zool. Soc. 1875.
   Cottyam Hills, S. India.

   We have seen only four specimens (none perfect) of this peculiar Glessula.

5. **A. Isis**, Hanley, Proc. Zool. Soc. 1875.
   Southern India.

   The hair-like lines and narrow fillet (the latter not a constant feature) are found in no other Glessula of so elongated a shape.

6. **Cor. Bengalensis**, Desh. Proc. Zool. Soc. 1854, p. 344; Cat. Brit. Mus. Vener. p. 221.—Prime, Ann. Lyc. N. Y. 1865, vol. 8, p. 220, f. 52.
   River Jumna.

7. **Cor. trigona**, Desh. Proc. Zool. Soc. 1854, p. 344; Cat. Brit. Mus. Vener. p. 221.—Prime, Ann. Lyc. N. Y. 1865, vol. 8, p. 221, f. 53.
   Pondicherry.

   This and the last are figured from the types in the British Museum.

8. **Cor. Iravadica**, Blanford, MSS.
   River Irawaddy, Pegu and Ava.

   This has been distributed by the author as pisum and iravadica. It has probably been published, but the reference cannot be found at the moment of going to press.

9. **Cy. Indica**, Deshayes, Proc. Zool. 1854, p. 342,

and Cat. Brit. Mus. Vener. p. 265 (as Sphæ-
rium).

Moradabad.

The original type was a little rounder and without
the coloured epidermis.

10. **P. Clarkeanum**, Nevill, J. Asi. Soc. Beng. 1871,
vol. 39, pt. 2, pl. 1, f. 4, 4 a, 4 d.

River Goomty; Moisraka, Dagunda, in tanks;
Chittagong.

Judging from Hutton's types, this is the Affghan
shell which he published, in English, as paludosum,
in the Asiatic Journal for 1849 (p. 659); but although his description is prior, it is not only in
English, but too meagre to ensure identification.

Prime has given the name solidulum to a supposed
P. porculum of Benson from "India." Can it be this
species?

PLATE CLVI.

ACHATINA, COILOSTELE, PUPA, STREPTAXIS.

See previous plates xvii, xviii, xxxv, xxxvi, lxxviii.
cii (for first); viii, xcviii (for last).

1. **A. Bottampotana**, Beddome, MSS.

Lent to us by Col. Beddome with this local name.

2. **A. hobes**, Blanford's MSS. in Pfeif. Mon. Hel.
vol. 6, p. 230 — Glessula p. Blanf. J. Asi. Soc.
Beng. 1870, vol. 39, p. 24, pl. 3, f. 21.

Deo Ghat, south of Poona.

3. **A. capillacea**, Pfeiffer, Proc. Zool. Soc. 1854,
p. 294; Mon. Helic. vol. 4, p. 614.

Ceylon.

The original type, now in the British Museum, is
here delineated: it is, however, in poor condition.

4. **A. Beddomei**, Blanford (see plate cii, f. 8).

Our previous figure copied from an Indian painting
represented certain accidental white specks as essen-
tial: our present is taken from the type, which differs
little from A. inornata, except in its peaked apex.

5. **C. scalaris**, Benson, An. Nat. H. 1864, Feb.

Found dead in the sand of the river Betwa, and
the left bank of the Jumna, and the exuviæ of
the Ganges.

6. **P. tutula**, Benson in Reeve, Conch. Icon. Bulim.

pl. 84, f. 625.—Pupa f. Pfeif. Mon. Helic. vol. 3,
p. 555, and Kust. Chemn. Pup. pl. 17, f. 8, 9, 10.

Hanseerpore, Bundelkhund.

Just intermediate between Bulimus and Pupa.

7, 8. **S. Canarica**, Beddome's name in Blanf. J.
Asi. Soc. Beng. vol. 38, pt. 2, 1869, pl. 16, f. 11;
Cont. Mal. pt. 10,

S. Canara, not far from west coast of India.

The type here delineated seems at present unique
(at least in England).

9. **S. bombax**, Benson (as Helix). See previous
plate 34, f. 1, 4.

This is the adult of the young shell we had
previously figured as a Helix.

10. **S. Birmanica**, Blanford, J. Asi. Soc. Beng.
1865, pt. 2, p. 84; Cont. Mal. pt. 5.—Pfeif. Mon.
Helic. vol. 5, p. 444—Stol. J. Asi. Soc. Beng.
1871, vol. 40, pt. 2, pl. 7, f. 5, 6, 7.

Tongoop, Arracan.

Either through transposition for symmetry's sake
by the lithographer after the printing of the text, or
misled by remarks in the Asiatic Journal, the names
of Birmanica and Blanfordi were interchanged in
part I of the Conchologia Indica.

PLATE CLVII.

NAVICELLA, NERITINA.

See, for Navicella, previous plate cxxxvii.

1, 4. **Nav. squamata**, Dohrn, Proc. Zool. Soc.
1858, p. 135.

Ceylon.

Drawn from the type in the British Museum.

2, 3. **Ner. Porcetiana**, Récluz, Rev. Zool. Cuv.
1841, p. 335. — Sow. Thes. Conch. vol. 2, pl. 115,
f. 200, 201.—Reeve, Conch. Icon. Ner. f. 124.

Ceylon; Nilgherries.

We have not obtained the Ceylonensis of Récluz
(J. Conc. 1854, p. 202) said to be very near this,
but with its black operule edged with orange.

5, 6. **Ner. reticularis**, Sowerby (for reticulata
Bens. not Sow. in Z. P.), Conch. Illus. Ner. f.
45; Thes. Conch. vol. 2, p. 536, f. 264, 265. —N.
humeralis, Th. (name only).

Bengal; Salwen.

The Salwen variety has been referred to as N. humeralis of Benson, but was never described.

7. **Ner. obtusa**, Benson, in Sow. Conch. Illust. Ner. f. 43; Thes. Conch. vol. 2, p. 517, pl. 111, f. 72, 73.—N. spiralis, Reeve, Conch. Icon. Ner. f. 40 (from type).

Tanks near Calcutta.

8, 9. **N. fuliginosa**, Theobald, Journ. Asi. Soc. Beng. 1859, vol. 27, p. 315.—N. reticularis, var. capillula, Sow. Thes. vol. 2, f. 265, 266.

Near Ava, Birmah.

The manuscript name, "cryptospira Benson," was applied to the young of this species by Mr. Theobald.

10. **N. colubor**, Thorp, MSS.

Ceylon.

Before adopting this manuscript name, the species has been vainly sought for in the respective monographs of Reeve and Sowerby.

PLATE CLVIII.

CAMPTOCERAS, LIMNÆA, SUCCINEA.

See previous plates lxix, lxx, for Limnæa, and lxvii, lxviii, for Succinea.

1, 2. **C. terebra**, Benson, Calcut. J. of N. H. 1842, p. 465.—Adams, Genera Shells, pl. 74, f. 1.—Blanf. J. Asi. Soc. Beng. 1871, vol. 40, pl. 2, pl. 2, f. 1.

Rampurga, Moradabad, Rohilkhand.

3, 4. **C. Austeni**, H. Blanford, J. Asi. Soc. Beng. 1871, vol. 40, pl. 2, p. 40, pl. 2, f. 2.

Nazirpur, not far from Shusong, Mymensing, Bengal.

5, 6. **C. lineata**, H. Blanford, J. Asi. Soc. Beng. 1871, vol. 40, pl. 2, p. 40, pl. 2, f. 3 (as Camptoceras).

With the last species.

7. **L. brevicauda**, Sowerby in Reeve's Conch. Icon. Limn. f. 105.

Cashmire (W. Blanford).

The locality assigned to this shell by Mr. Sowerby is Australia: the identity of his type with the unique specimen here figured was proved by careful comparison. It will by some—and perhaps rightly—be regarded as a form of L. auricularia which is found

in the same parts; yet its peculiar spire, its more globose shape, and the spirally tortuous columellar fold afford distinguishable characters.

9. **S. subgranosa**, Pfeiffer, Proc. Zool. Soc. 1849, p. 132; Mon. Helic. vol. 3, p. 9.

Kurnaal, and near Calcutta (a variety).

10. **S. Ceylanica**, Pfeiffer, Proc. Zool. Soc. 1854, p. 297; Mon. Helic. vol. 4, p. 810.

Ceylon.

PLATE CLIX.

HELIX, ACHATINA, CLAUSILIA.

See previous plates xiii to xvi, xxv to xxxii, l to lxiv, lxxxiii to xc, cxi, cxii, cxxvii to cxxxii, cxlix, cl, first; xvii, xviii, xxxv, xxxvi, lxxviii, cii, clv, for second; xxiv, cxviii, for third.

1, 4. **H. biciliata**, Pfeiffer, Proc. Zool. Soc. 1855, p. 112; Mon. Helic. vol. 4, p. 68.

Ceylon.

2, 3. **H. infausta**, Blanford, J. Asi. Soc. Beng. 1866, vol. 34, p. 36, and Cont. pt. 6, as Nanina.—Pf. Mon. Helic. vol. 5, p. 124.

Anamallay Hills, S. India.

5. **C. Asaluensis**, Godwin-Austen, MS. in Blanf. J. Asi. Soc. Beng. 1871, vol. 41, pt. 2, p. 292, pl. 9, f. 8.

Asalu, North Cachar.

6. **A. orthoceras**, Godwin-Austen, J. Asi. Soc. Beng. 1875, vol. 44, pt. 2, p. 2, pl. 1, f. 1.

W. Khasi Hills.

Whorls more oblique and more rounded than in Cassiaca: no epidermis.

7. **H. Shiroiensis**, Godwin-Austen, Proc. Zool. Soc. 1874, p. 609, pl. 75, f. 3.

8. **H. confinis**, Blanford, J. Asi. Soc. Beng. 1865, pt. 2, p. 71; Cont. Mal. Ind. pt. 5.—Pf. Mon. Helic. vol. 5, p. 89.

Thayet Mio, confines of British Birmah.

9. **H. Humberti**, Brot, J. Conch. 1864 (vol. 12), p. 21, pl. 2, f. 5, 6.—Pf. Mon. Hel. vol. 5, p. 398.

Ceylon.

Only one palatal and one parietal tooth. The front might pass for odontophora, but the lower disk is more smooth, and just like our figure of erronea. The last-named species very clearly represented in the same plate of the Journal de Conchyliologie has three parietal and three palatal lamellæ visible at the mouth, but a fourth palatal lamella may be traced by the opaque marks on the exterior. This does not quite suit Alber's original (Zeit. Mal. 1853, p. 305) description, and we possess a variety (?) with coarser sculpture and a lesser number of ridges which agrees better.

10. **H. Beddomei**, Blanford, Ann. Nat. Hist. 1874, ser. 4, vol. 14, p. 406 (as Hemiplecta).

Western side of Travancore Hills, S. India.

Equally allied to basileus and Chenui.

PLATE CLX.

PUPA.

See previous plates c, ci, clvi.

1. **P. (Ennea) sculpta**, Blanford, J. Asi. Soc. Beng. 1869, vol. 38, p. 141, pl. 16, f. 10.
   Pulney Hills, Southern India.

2. **P. muscerda**, Benson, Ann. Nat. Hist. 1857, ser. 2, vol. 12, p. 94.—Pfeif. Mon. Helic. vol. 4, p. 680.
   On old posts and Palmyra trees, Pedro Promontory, Ceylon.

3. **P. plunguncula**, Benson. See previous plate 101, f. 2.

Having at last obtained a perfect specimen of this rare shell from the Kunah Hills in Southern India, we have again delineated it.

4. **P. mimula**, Benson. Ann. Nat. Hist. 1853, ser. 2, vol. 12, p. 95.—Pfeif. Mon. Helic. vol. 4, p. 676.
   In vines on Pedro Promontory, Ceylon.

5. **P. filosa**, Theobald and Stoliczka, J. Asi. Soc. Beng. 1874, vol. 41, pt. 2, p. 385, pl. 11. f. 8.
   Aracan Hills.

6. **P. lignicola**, Stoliczka, J. Asi. Soc. Beng. 1871, vol. 40, pt. 2, p. 171, pl. 7, f. 3.
   Moulmein.

We have figured the toothed variety.

7. **P. Avanica**, Benson, An. Nat. Hist. 1853, p. 428.—Pfeif. Mon. Helic. vol. 6, p. 335.
   Ava, Birmah.

8. **P. Indica**, Pfeiffer, Proc. Zool. Soc. 1854, p. 295 ; Mon. Hel. vol. 4, p. 679.
   Barrakpore, near Calcutta.

This is not the Indica of Benson (name only for the P. cylindrica of Hutton) in the J. Asi. Soc. Beng. vol. 18, pt. 2, p. 673, quoted by Pfeiffer as equal to the Bulimus pullus. The types here represented are in the National Museum.

9. **P. Salemensis**, W. and H. Blanford, J. Asi. Soc. Beng. 1861, vol. 30, p. 359, pl. 2, f. 8 : Cont. Mal. pt. 2.— Pfeif. Mon. Hel. vol. 6, p. 318.
   Kalryen Hills, S. India.

We have merely copied the original figure.

10. **P. (Ennea) cylindroides**, Stoliczka, J. Asi. Soc. Beng. 1871, vol. 40, pt. 2, p. 171, pl. 7, f. 4.
    Damolha, near Moulmein.

Copied from the cited figure, for want of a specimen

Plate 1

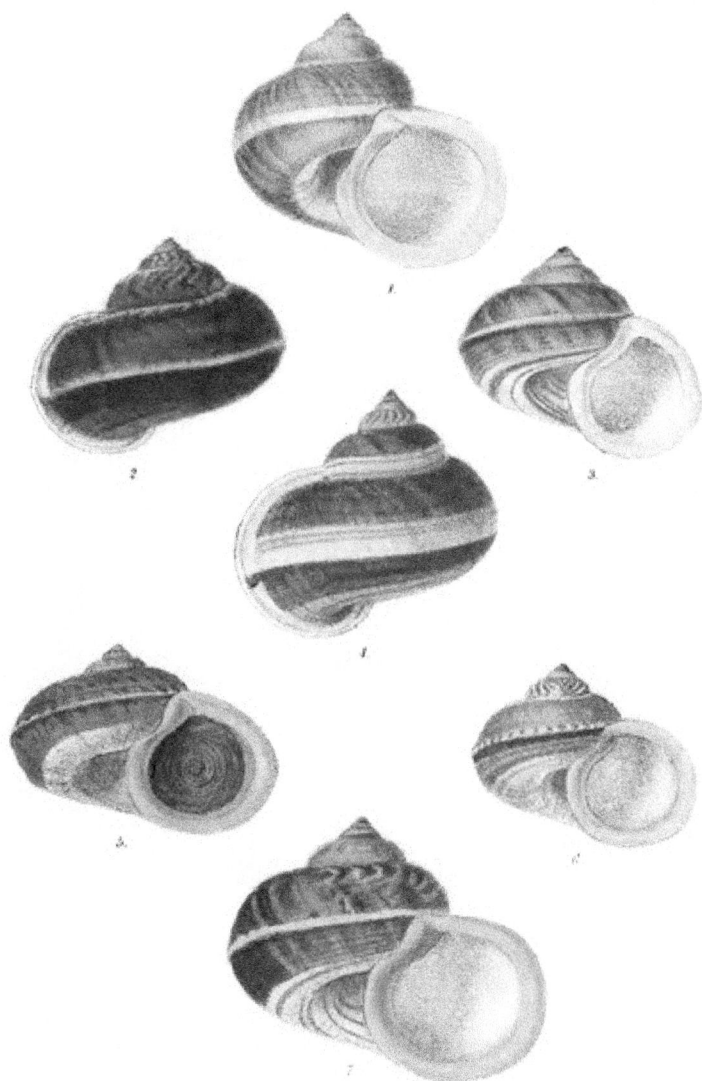

G B Sowerby del et lith.

Vincent. Brooks Davidson imp

L. Reeve & Cº Publishers S Henrietta st Covent Garden

Plate 11

1

2

3

4

5

6

7

8

9

10

G B Sowerby, del et lith   Vincent Brooks Day & Son Imp

L Reeve & Co Publishers 5 Henrietta St Covent Garden.

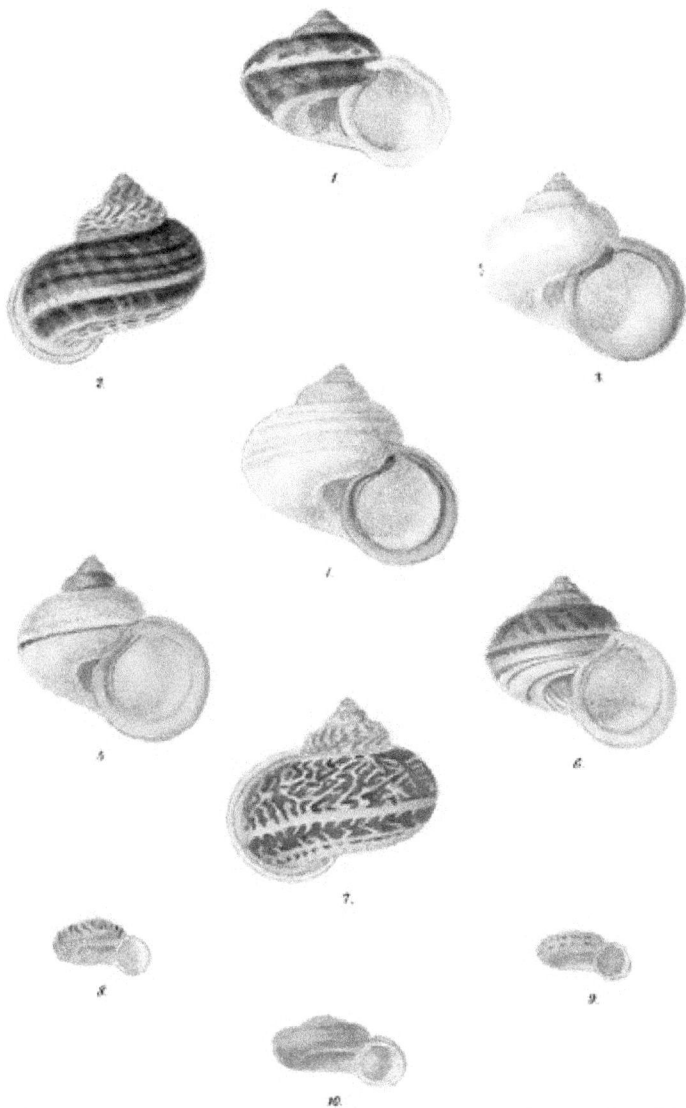

1.

2.

3.

4.

5.

6.

7.

8.

9.

10.

G. B. Sowerby, del. et lith.

Vincent Brooks, Day & Son, Imp.

L. Reeve & Co. Publishers, 5 Henrietta St. Covent Garden.

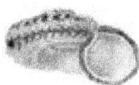

G B Sowerby del et lith

Vincent Brooks Day & Son, Imp

L. Reeve & Cº Publishers 5 Henrietta St Covent Garden

Plate V

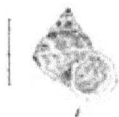

1

2

4

5

6

7

10

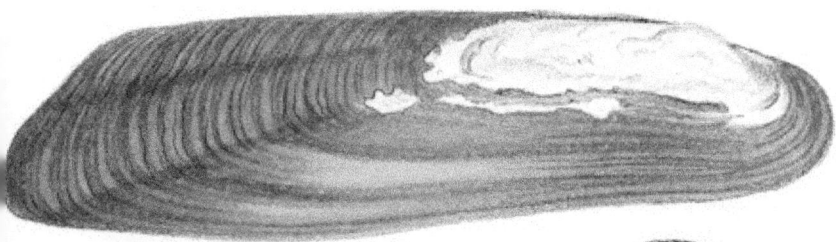

1.

3.

2.

4.

5.

6.

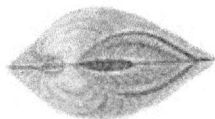

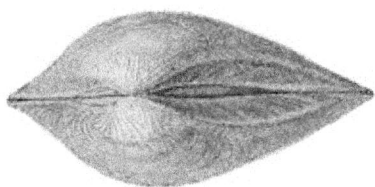

1

2

3

4

4

5

6

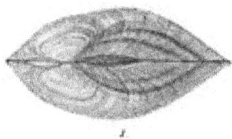

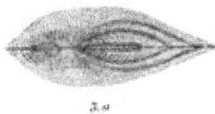

G B Sowerby del et lith

Vincent Brooks Day & Son Imp

L Reeve & Co Publishers 5 Henrietta St Covent Garden

2

3

4

5

6

7.

8

9

10.

G B Sowerby, del et lith.

Vincent Brooks Day & Son Imp

I. Reeve & Co Publishers 5. Henrietta St Covent Garden

_1_

_2._

_3._

_4_

_5._

_6._

_7_

_8_

_9_

_10._

G B Sowerby del. et lith

Vincent Brooks Day & Son, Imp

L Reeve & Co. Publishers, 5 Henrietta St Covent Garden

G B Sowerby del et lith.

Vincent Brooks Day & Son. imp

1. Reeve & Co Publishers 5. Henrietta St Covent Garden.

Vincent Brooks Day & Son lith

L. Reeve & Co. Publishers, 5, Henrietta St Covent Garden

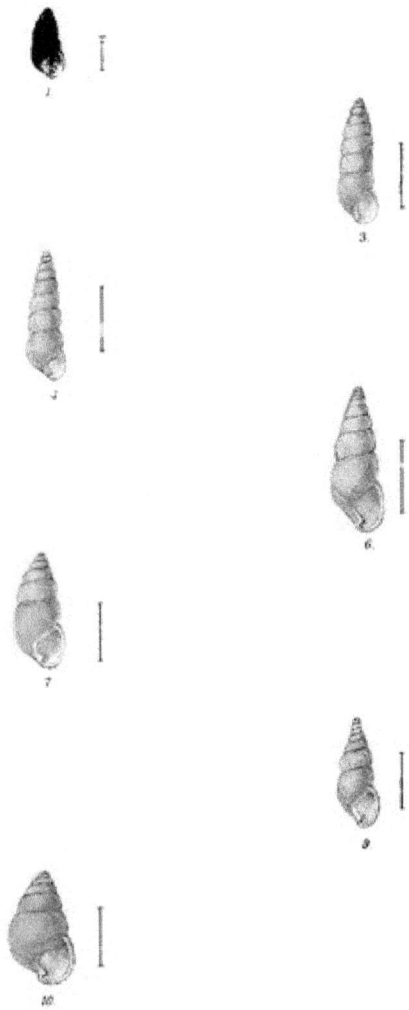

Vincent Brooks Day & Son, Imp

L Reeve & Co Publishers 5 Henrietta St Covent Garden

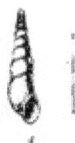

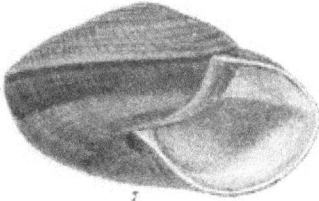

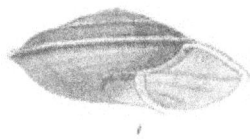

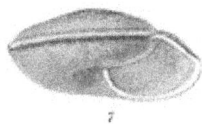

G B Sowerby, del et. lith.

Vincent Brooks Day & Son Imp

G.B. Sowerby del. et lith.   Vincent Brooks Day & Son. Imp

L. Reeve & Co. Publishers, 5 Henrietta St. Covent Garden.

2

3

4

5

6

7

8

9

10

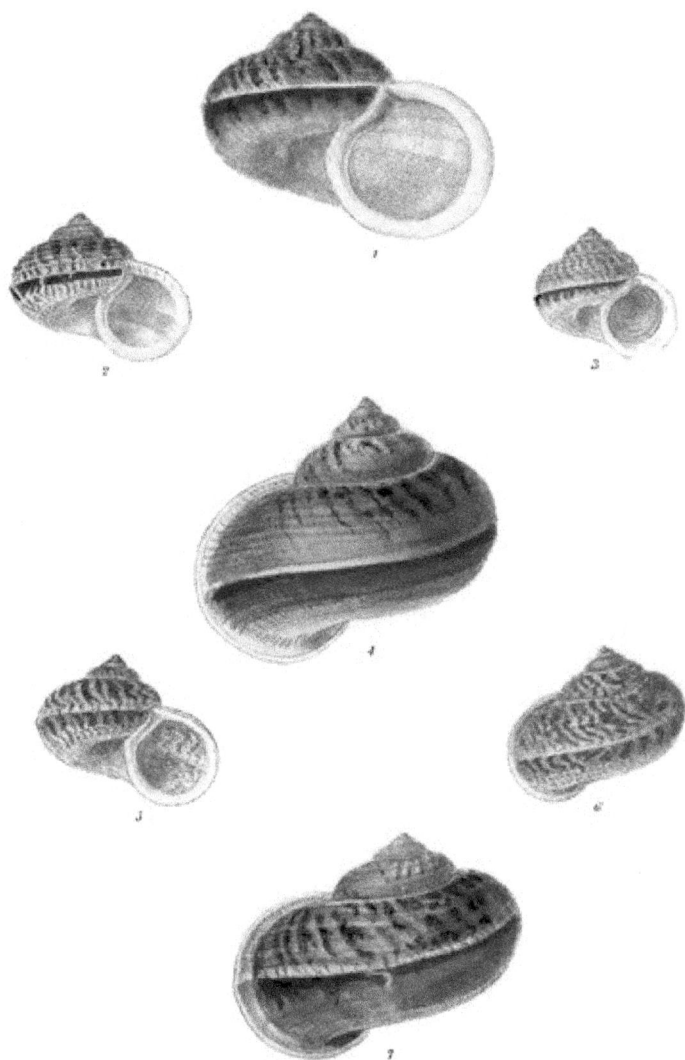

G B Sowerby, del et lith.

Vincent Brooks, Day & Son Imp

L Reeve & Co Publishers, 5, Henrietta St Covent Garden

G B Sowerby, del et lith

Vincent Brooks Day&Son. Imp.

L. Reeve & Co Publishers,5. Henrietta St Covent Garden

2

3

Plate XXXVII

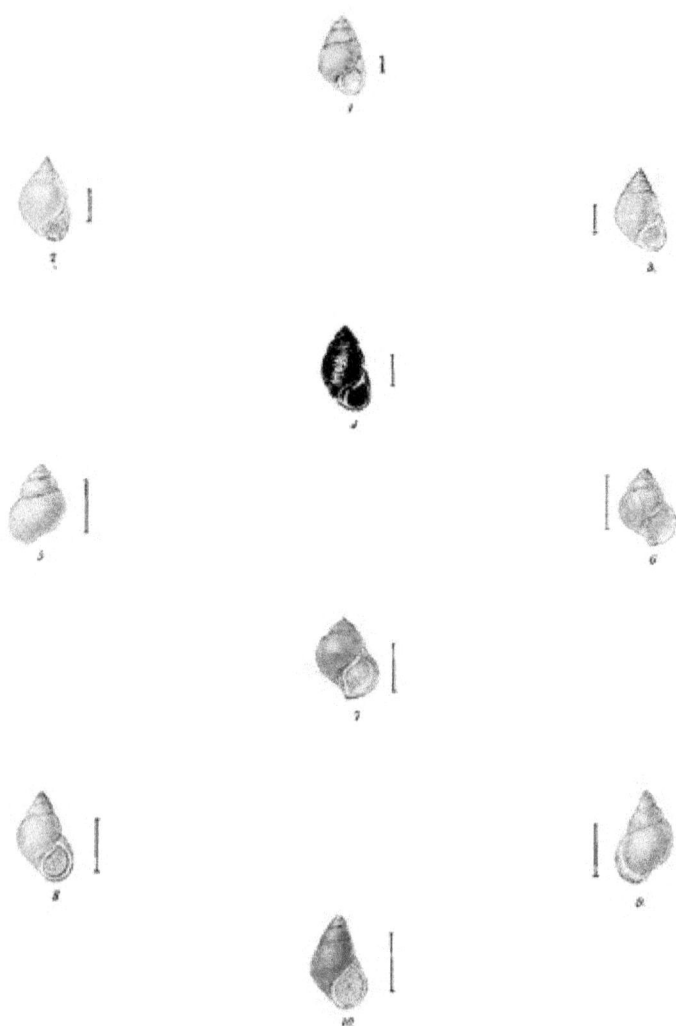

G B Sowerby del et lith.

Vincent Brooks Day & Son. Imp

L Reeve & Co Publishers, 5, Henrietta St Covent Garden

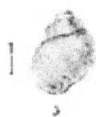

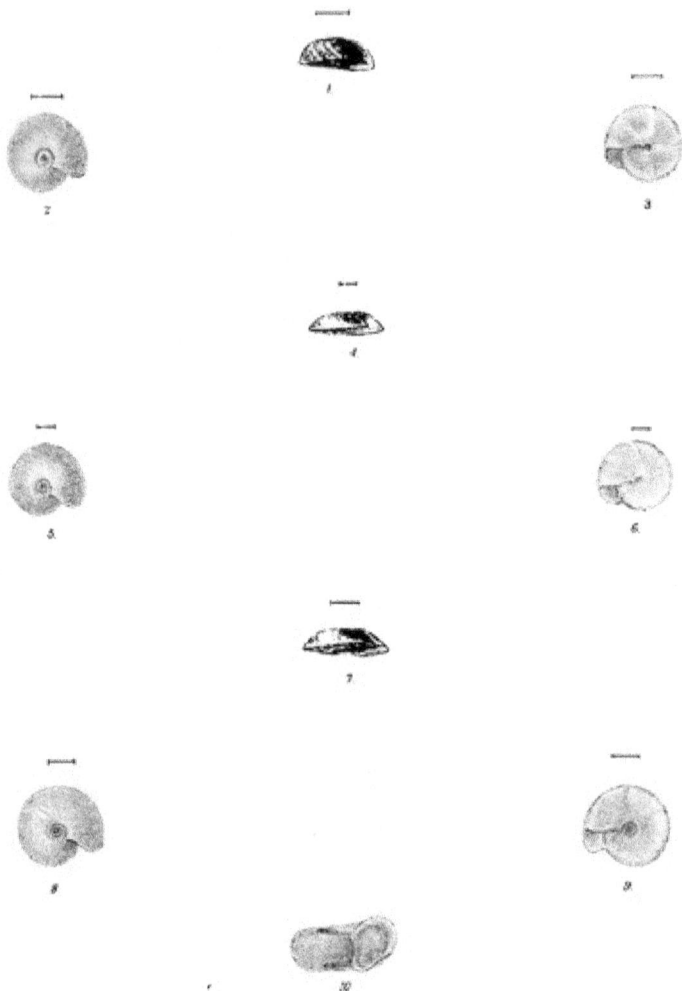

G.B. Sowerby del et lith.

Vincent Brooks Day & Son Imp.

L. Reeve & Co Publishers 5 Henrietta St Covent Garden

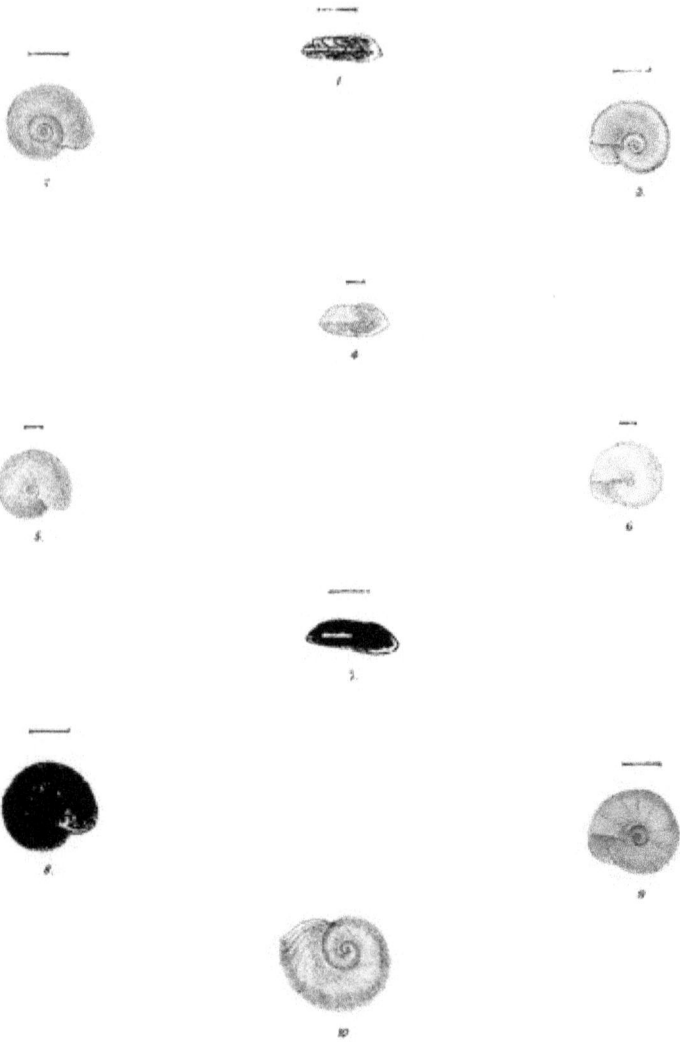

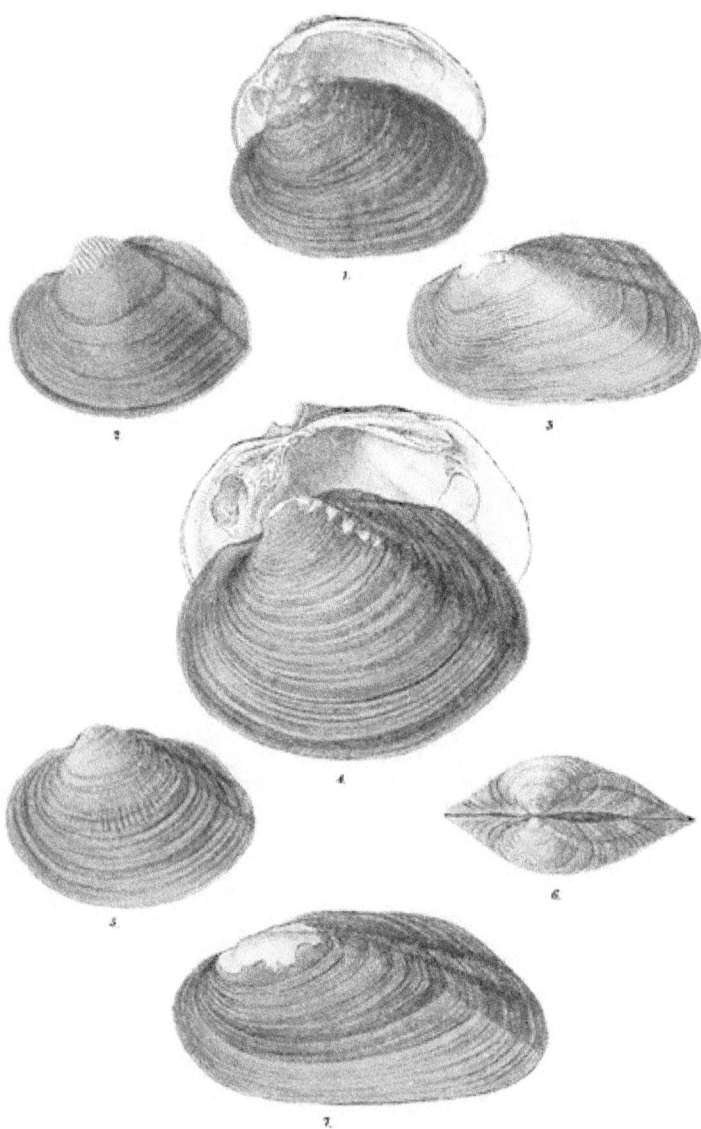

1.

2.

3.

4.

5.

6.

7.

L. Reeve & Co Publishers, 5, Henrietta St. Covent Garden.

Plate XLVI

4

3

4

2

1.

2.

3.

4.

6.

7.

5.

8.

9.

10.

G.B.Sowerby, del et lith.

Vincent Brooks Day & Son Imp.

L. Reeve & Co Publishers, 5, Henrietta Covent Garden.

1.

2

3.

4.

5.

6.

7.

8.

9

10.

G.B.Sowerby. del et lith.

Reeve Broth. imp

L. Reeve & Co. Publishers, b. Henrietta st. Covent garden.

2

3

4

5

6

7

8

9

10

G.B. Sowerby del. et lith.   Hanhart Imp.

1.

3.

4.

2.

5.

7.

6.

8.

G. B. Sowerby, del et lith.

Vincent Brooks Day & Son, Imp.

L. Reeve & Co. Publishers, 5, Henrietta St. Covent Garden

1.

2.

3.

5.

6.

4.

7.

8.

3

5

4

5

6

7

8

9

10

J. Reeve & Co. Publishers to Heralorica & Covent Garden

4

6

7

1

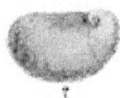

2

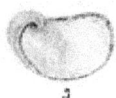

3

4

5

6

7

8

9

10

1

2

3

4

5

6

8

9

7

10

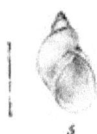

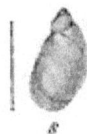

1

2

4

6

7

9

10

Mawest Brooks, Day & Son, Imp.

Reeve & Henrietta St Covent Garden.

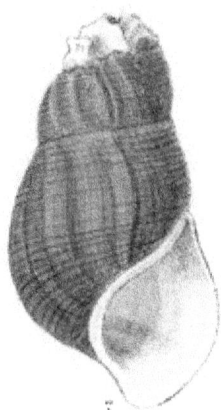

Pl i XXX

1

3

4

6

7

9

10

Vincent Brooks Day & Son Imp

L. Reeve & Co Publishers 5 Henrietta St Covent Garden

G. B. Sowerby del. et lith.   Vincent Brooks Day & Son imp.

L. Reeve & Co. Publishers, 5 Henrietta St. Covent Garden.

1.

3

4

6

7

9

10

L. Reeve & Co. Publishers, 5 Henrietta St. Covent Garden.

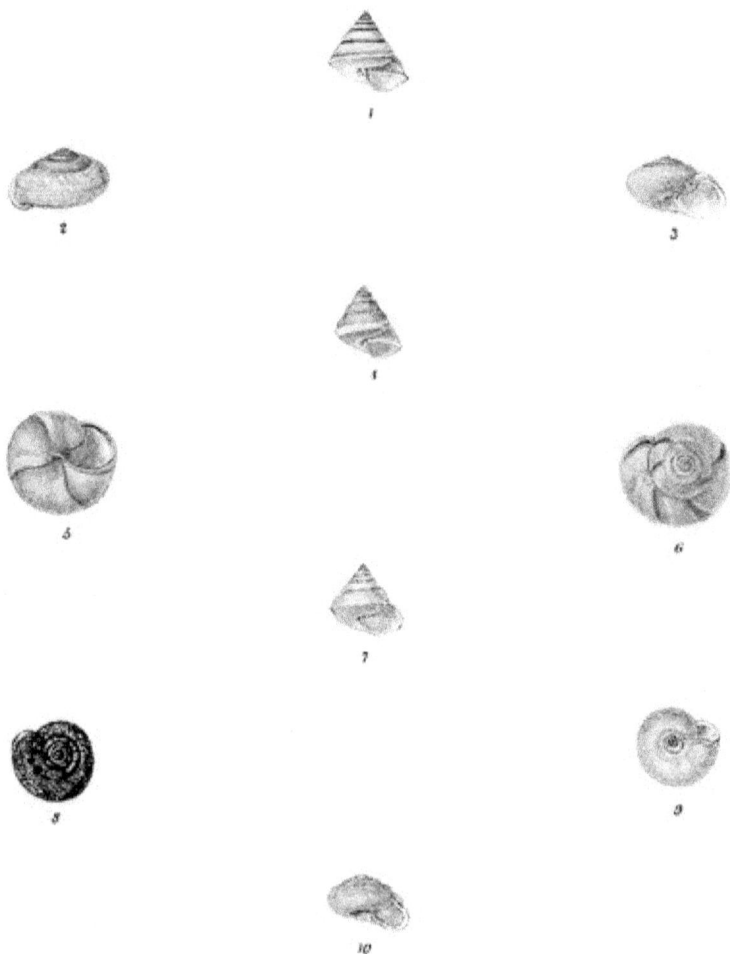

1

2   3

4

5   6

7

8   9

10

J. B. Sowerby del. et lith.   Vincent Brooks Dayl. & Son Imp.

L. Reeve & Co. Publishers, 5 Henrietta St. Covent Garden.

Pl. LXXXVII.

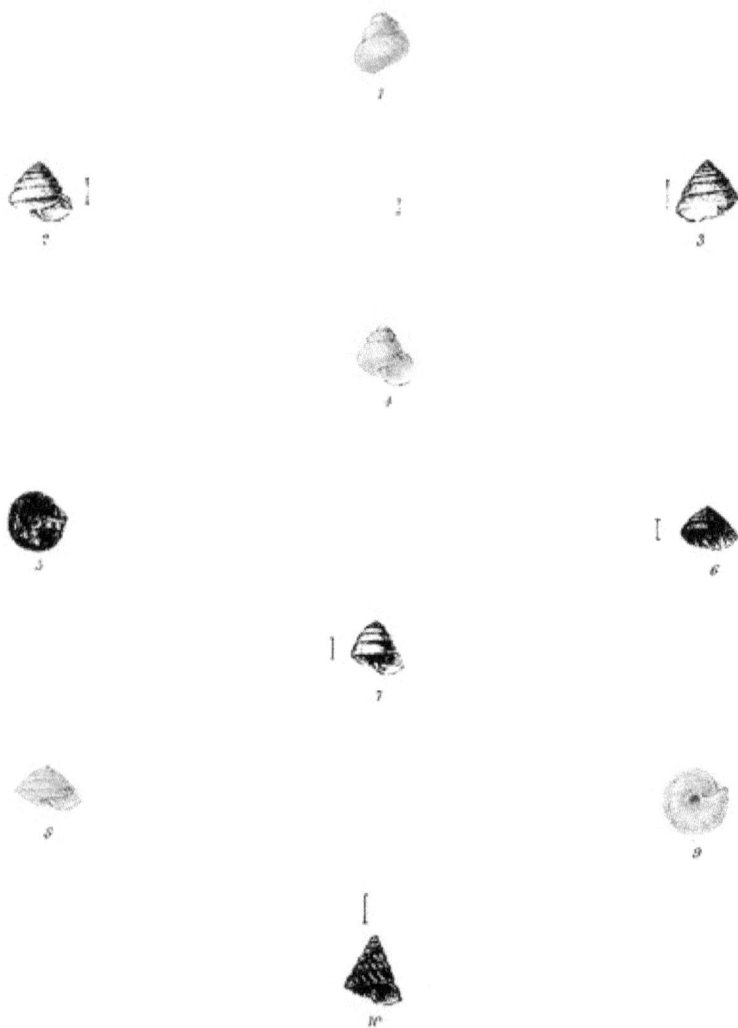

1

2

3

4

6

5

7

8

9

10

G. B. Sowerby del et lith.

Vincent Brooks Day & Son Imp.

L. Reeve & Co. Publishers, 5 Henrietta St. Covent Garden.

2

3

4

5

6

7

8

9

10

1

2

3

4

5

6

7

8

9

10

C.B.Sowerby del et lith.

Vincent Brooks Day & Son Imp.

L. Reeve & Co. Publishers 5 Henrietta St Covent Garden.

6

3

2

5

5

6

7

9

8

Vincent Brooks Day & Son, Imp.

L. Reeve & Cº Publishers: 5 Henrietta Sº Covent Garden.

2

3

1

5

6

7

8

9

10

G.B. Sowerby del et lith.   Vincent Brooks Day & Son imp.

L. Reeve & C? Publishers, 5 Henrietta S? Covent Garden.

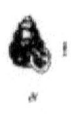

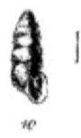

1

3

4

5

6

7

9

10

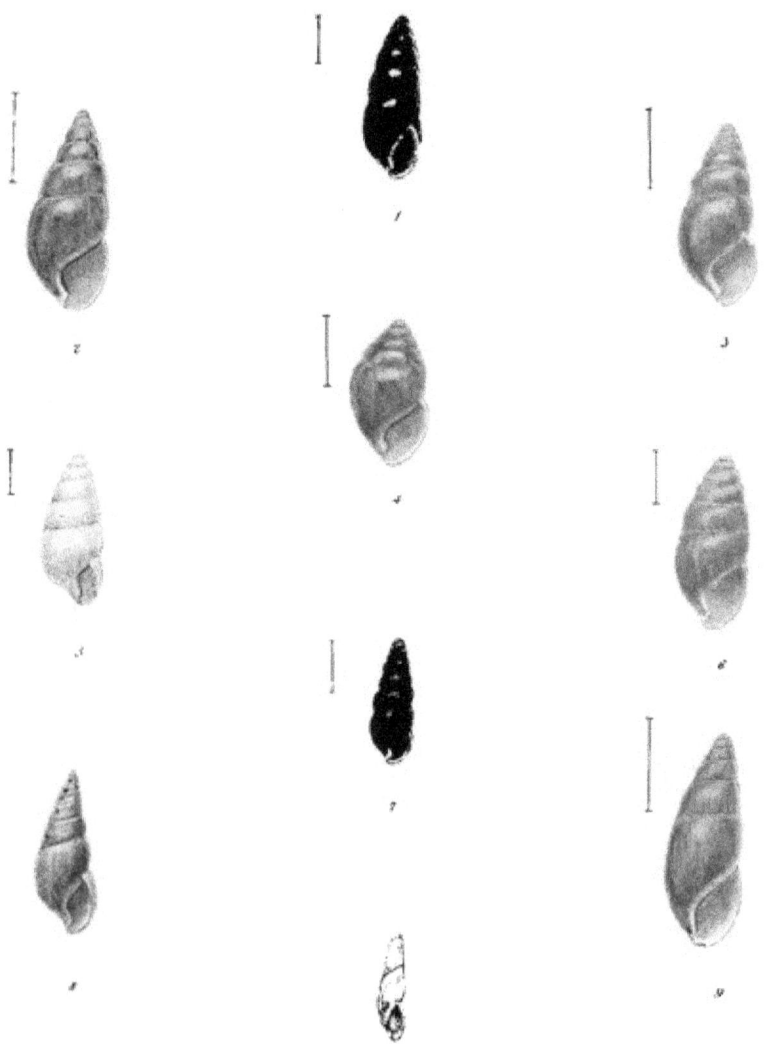

Carl von Bouell, del et hth

L. Reeve & Co Publishers Henrietta St Covent Garden

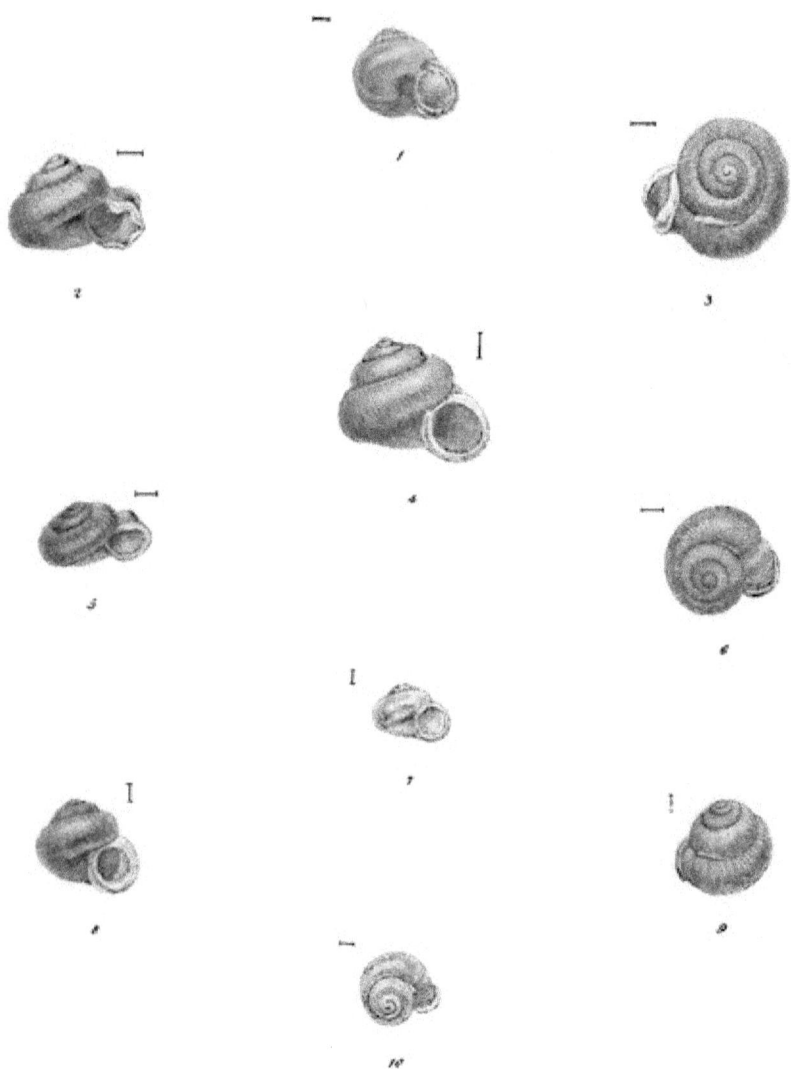

Carl von Bouell. del et lith

Photo Litho Inst. Imp.

J. Reeve & Cº Publishers, 5 Henrietta St. Covent Garden.

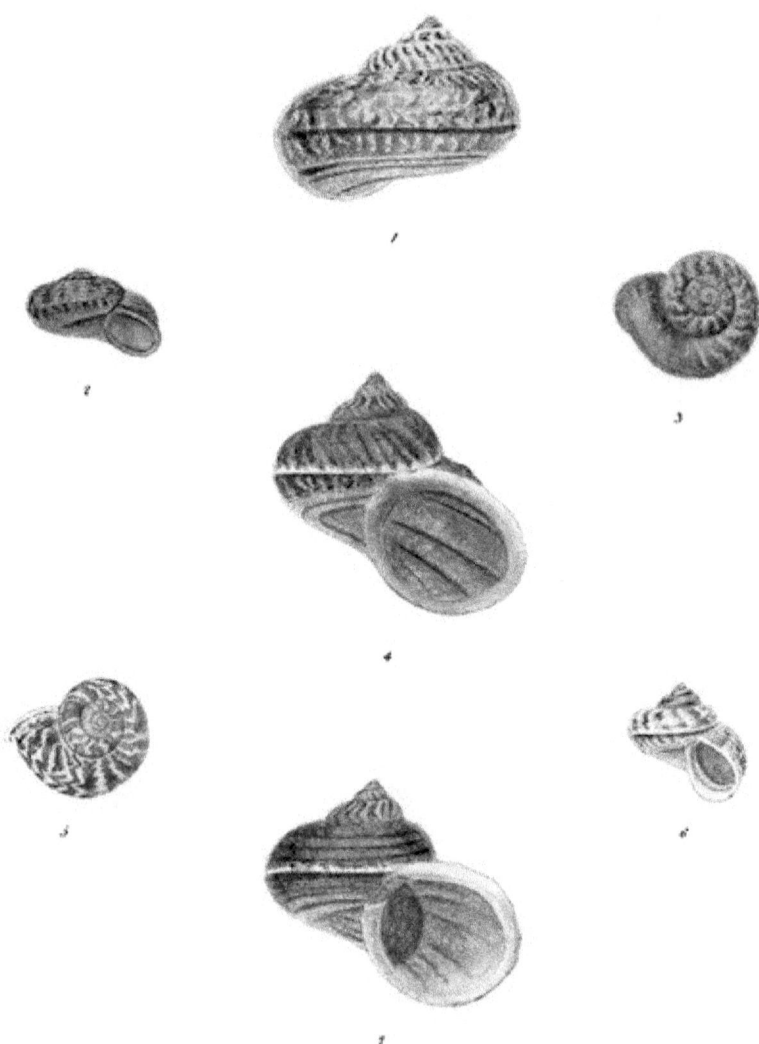

Carl von Bouell del et lith.

Photo Lith. Inst. Imp.

L. Reeve & Co Publishers 5 Henrietta St. Covent Garden.

Plate CXIII

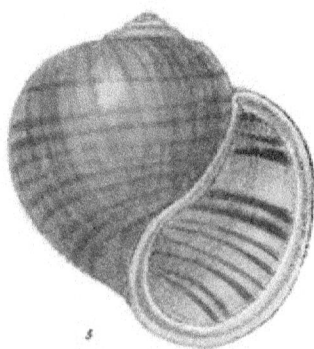

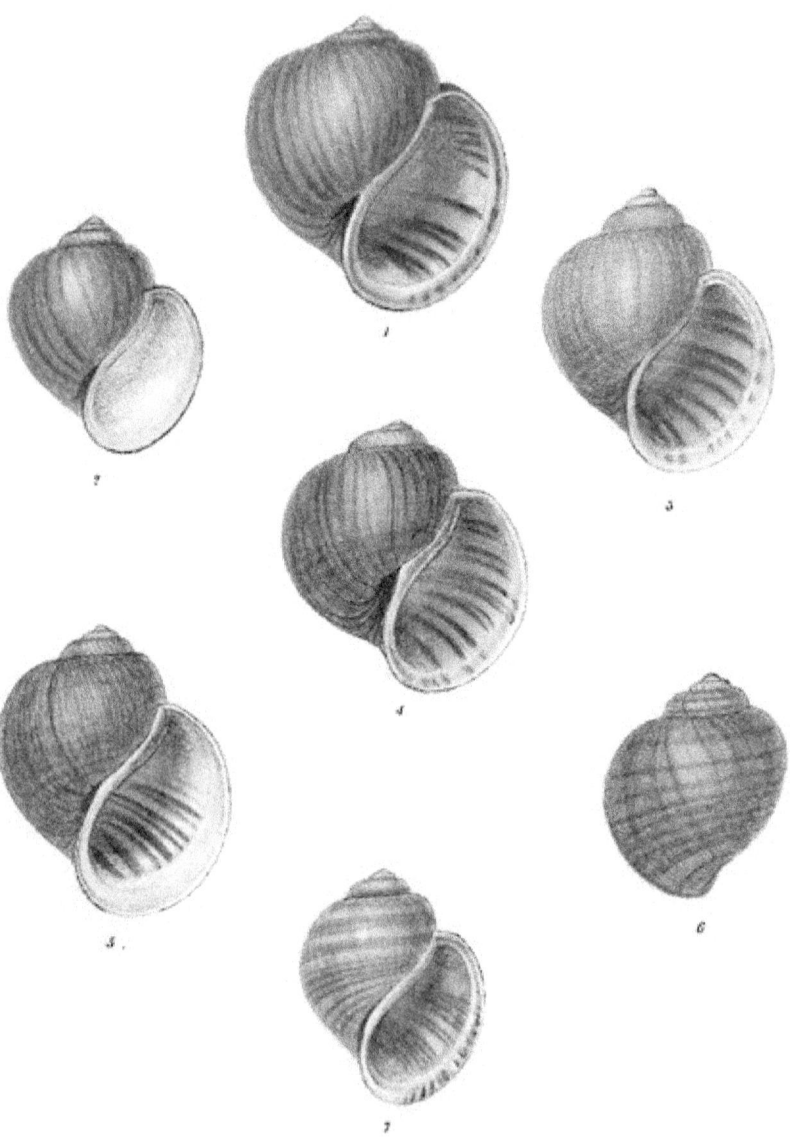

1

2

3

4

5.

6

7

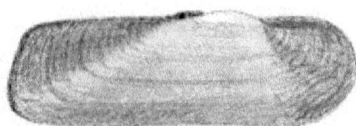

2

3

4

5

6

7.

8

9

10

All rights reserved

1

2

3

4

5

6

7

8

9

10

1

2

3

4

5

6

7

8

10

9

1

2

3

4

5

6

7

8

9

10

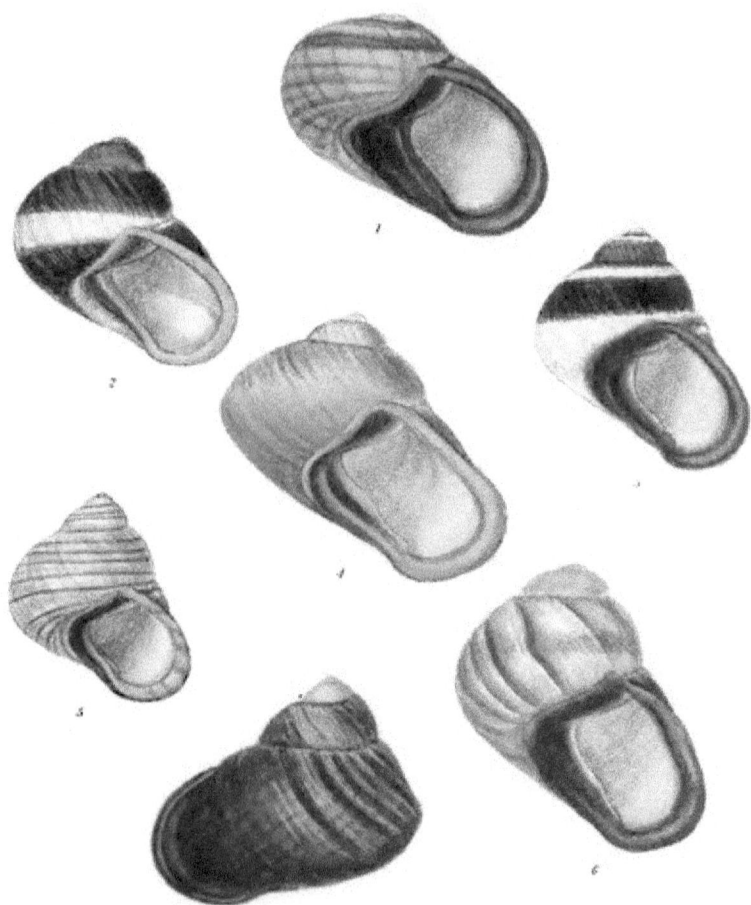

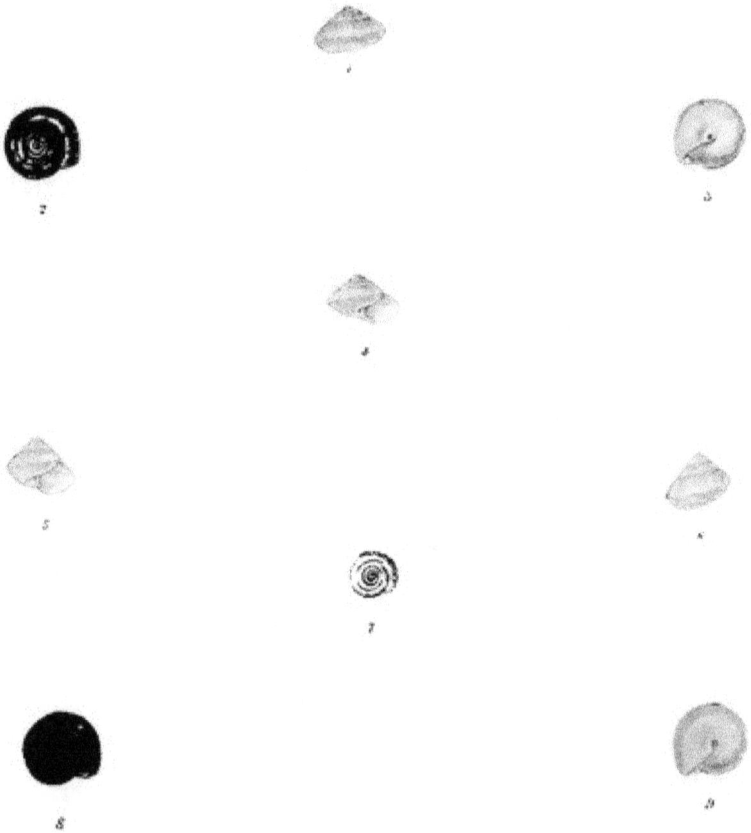
2
5
7
8

2

3

4

6

7

9

10

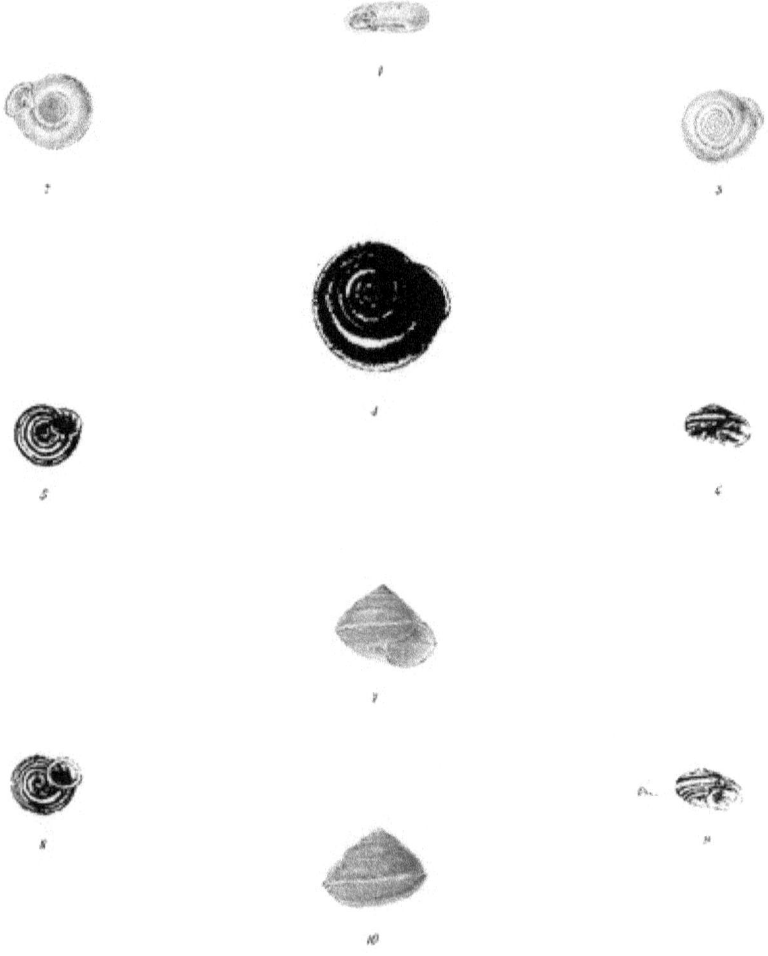

[illegible] Vincent Brooks Day & Son Imp.

L. Reeve & C° Publishers, 5, Henrietta St Covent Garden.

1

2

3

4

5

6

7

8

9

10

4

5

6

7

8

9

1

2

3

4

5

6

7

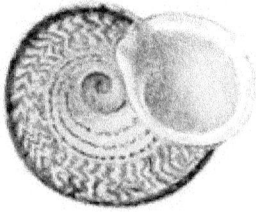

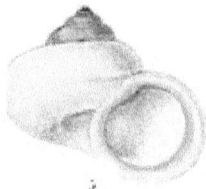

3.

1.

4.

5.

7.

8.

10.

1.

2.

3.

4.

5.

6.

7.

8.

9.

10.

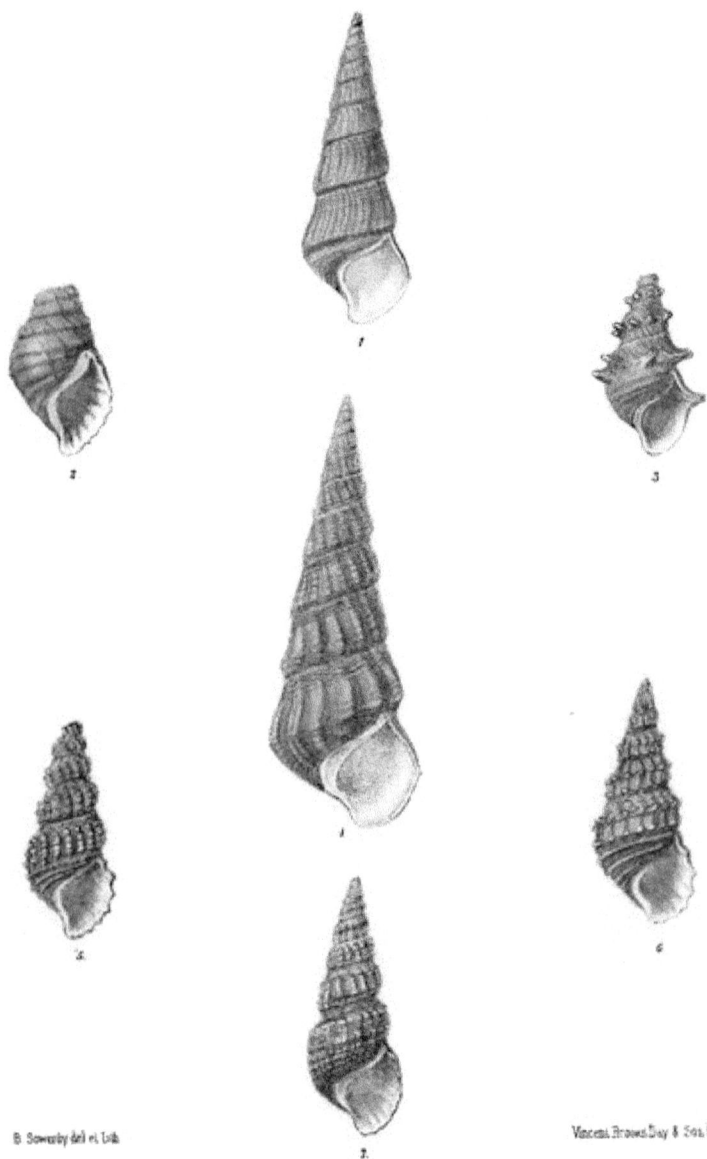

1

2

3

4

5

6

7

B. Sowerby del. et lith.

Vincent Brooks Day & Son. l.

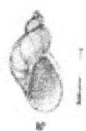

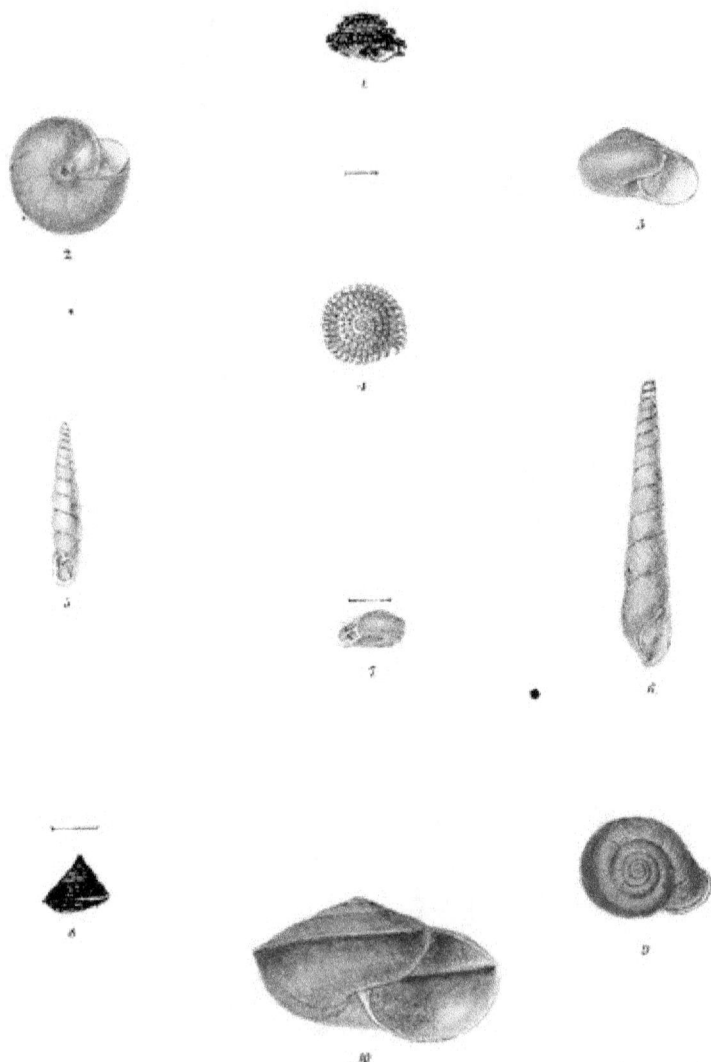

www.ingramcontent.com/pod-product-compliance
Lightning Source LLC
Chambersburg PA
CBHW021354210326
41599CB00011B/872